Holt
Mathematics
Course 1

Texas Lab
Manual Workbook

ISBN 0-03-092715-3

4 5   018   09 08 07

Table of Contents

Holt Mathematics

Holt Mathematics

Holt Mathematics

Holt Mathematics

Name _____ Date _____ Class _____

TEKS 6.13.B.02

 LAB 1-4 # Validate Conclusions When Simplifying an Expression

Use with Lesson 1-4

Materials needed: pencil
Activity
The order of operations tells you the steps you should use when you simplify an expression.

---

1. Perform operations in parentheses.

2. Find the values of numbers with exponents.

3. Multiply or divide from left to right as ordered in the problem.

4. Add or subtract from left to right as ordered in the problem.

Choose a reason from the list on the right to justify each step in simplifying the following expression.

$1 + (6 - 2) \times 3^2 \div 2$

1. $1 + 4 \times 3^2 \div 2$ _____

2. $1 + 4 \times 9 \div 2$ _____

3. $1 + 36 \div 2$ _____

4. $1 + 18$ _____

5. 19 _____

a. Add.

b. Perform operations within parentheses.

c. Divide.

d. Multiply.

e. Find the values of numbers with exponents.

Try This
Simplify each expression. Justify each step using the order of operations.

1. $6 \times (2 + 8) \div 5 - 1$

_____

_____

2. $2^3 \times (5 - 3) \div 4$

_____

_____

Holt Mathematics

TEKS 6.13.B.02

 Validate Conclusions When Simplifying an Expression

3. $(4 \times 2) \div 2^2 + 7$

_____

_____

4. $5 + (12 - 4)^2 \div 16$

_____

_____

Holt Mathematics

TEKS 6.11.D

Choose the Method of Computation

Use with Lesson 1-6

Activity

Pair up with a classmate. Write down your age in years and months, your height in feet and inches, and how much money you have with you in dollars and cents. If you don't know your height, use a yardstick or tape measure to find it.

Record these numbers below.

Student 1   **Student 2**

Age: Years _____ Months _____   Age: Years _____ Months _____

Height: Feet _____ Inches _____   Height: Feet _____ Inches _____

Money: Dollars _____ Cents _____   Money: Dollars _____ Cents _____

Think and Discuss

1. Finding the difference between the two ages would be easier if you first converted each age from years and months to months only. How could you do this?

   _____

   _____

2. Which solution method (pencil and paper, mental math, or calculator) would you use to convert the ages to months? Explain your choice.

   _____

   _____

3. Explain how you could compute your age in days.

   _____

   _____

Holt Mathematics

★ **LAB 1-6** **Choose the Method of Computation**

4. Which solution method (pencil and paper, mental math, or calculator) would you use to find your age in days? Explain your choice.

_____

_____

Try This

1. Compute your age in months, and then compute your classmate's age in months. Find the difference.

_____

_____

2. Choose a solution method and find the difference in your heights in inches. Explain your solution method choice.

_____

_____

_____

3. Choose a solution method and find the total amount of money in cents you both have. Explain your solution method choice.

_____

_____

Holt Mathematics

TEKS 6.13.A

Examples and Nonexamples of Geometric Sequences

LAB 1-7A

Use with Lesson 1-7

Materials needed: pencil
Activity
In this lab, you will explore geometric sequences. The chart shows examples of sequences that are geometric sequences and examples of sequences that are not geometric sequences.

Geometric Sequences	Not Geometric Sequences
2, 4, 8, 16, . . .	2, 4, 6, 8, . . .
5, 15, 45, 135, . . .	10, 11, 12, 13, . . .
1, 4, 16, 64, . . .	1, 3, 6, 10, ...
2, 20, 200, 2000, . . .	4, 7, 5, 8, 6, 9, . . .

1. What do all of the geometric sequences have in common?

_____

2. What do you think is the definition of a geometric sequence?

_____

_____

Try This
Decide whether each sequence is a geometric sequence. If so, describe the pattern and give the next term.

1. 3, 6, 12, 24, ...

_____

2. 5, 7, 9, 11, ...

_____

3. 1, 3, 9, 27, ...

_____

4. 20, 16, 12, 8, ...

_____

Holt Mathematics

Name _____ Date _____ Class _____

  LAB 1-7B # Patterns in Pascal's Triangle

Use with Lesson 1-7

Pascal's Triangle is used in a variety of mathematical problems. In this activity, it is used to explore patterns.

Activity

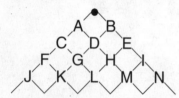

Find the number of paths from the top circle to each letter. You cannot move up when finding a path.

A   There is only one path from the top to A.
B   There is only one path from the top to B.
C   There is only one path from the top to C (through A).
D   There are two paths to D (one through A and one through B).
E   There is only one path from the top to E (through B).
F   There is only one path to F (through A and C).
G   There are three paths to G (one through A and C, one through A and D, and one through B and D).

Continue finding the number of paths for the other letters.

1. H _____   2. I _____
3. J _____   4. K _____
5. L _____   6. M _____
7. N _____

8. Fill in the triangle below with the number of paths to each point based on your answers from the activity above.

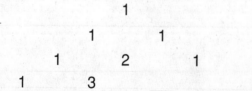

Holt Mathematics

TEKS 6.13.A

Patterns in Pascal's Triangle

Think and Discuss

1. The triangle shown in Exercise 8 on page L6 is referred to as Pascal's Triangle. What patterns do you observe in its numbers?

   _____

   _____

2. How could you find the numbers in the next line of the pattern without counting paths?

   _____

   _____

Try This

1. Write the numbers in the next two lines of Pascal's Triangle.

       ___   10   ___   ___   ___

   ___   ___   ___   ___   ___   6   ___

2. Find the sum of each row in Pascal's Triangle. Begin with the top "1."

   1st Row: _____

   2nd Row: _____

   3rd Row: _____

   4th Row: _____

   5th Row: _____

   6th Row: _____

   7th Row: _____

3. Using your answer from Exercise 2 above, what pattern do you observe in the sum of the rows? What would be the sum of the numbers in the 8th row?

   _____

   _____

Holt Mathematics

TEKS 6.11.D

Solving Addition Equations

Use with Lesson 2-5

You can use algebra tiles to help you solve equations.

KEY

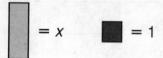

= x   ■ = 1

Activity

To solve the equation $x + 1 = 4$, you need x by itself on one side of the equal sign. Show this with algebra tiles. You can add or remove tiles as long as you add the same amount or remove the same amount on both sides.

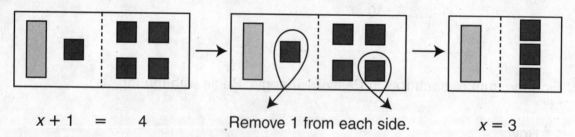

$x + 1 = 4$ Remove 1 from each side. $x = 3$

1. Circle the tiles to remove to solve $x + 3 = 5$. Then draw the tiles in the last box.

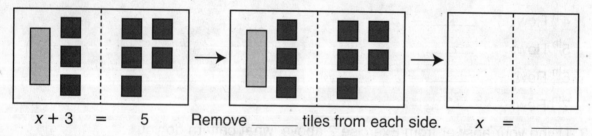

$x + 3 = 5$ Remove _____ tiles from each side. $x = $ _____

Holt Mathematics

TEKS 6.11.D

LAB
2-5

Solving Addition Equations

Think and Discuss

1. When you remove tiles, what operation are you representing?

_____

2. Give an example of an equation that would require you to remove four tiles to solve.

_____

Try This

Draw algebra tiles to model each equation. Then solve the equation.

1. $x + 2 = 7$

$x =$ _____

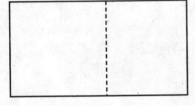

2. $x + 1 = 5$

$x =$ _____

3. $x + 5 = 8$

$x =$ _____

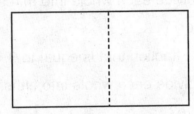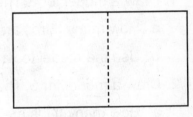

4. $3 + x = 4$

$x =$ _____

5. $x + 4 = 9$

$x =$ _____

6. $2 + x = 2$

$x =$ _____

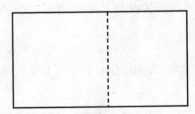

Holt Mathematics

TEKS 6.1.B.01

LAB 4-5

Whole Numbers and Fractions

Use with Lesson 4-5

Materials needed: pencil
Activity

You can use drawings to model the connection between whole numbers and fractions. For example, the whole number 3 is modeled by three squares. Each square is divided into fourths.

There are 12 fourths altogether so $3 = \frac{12}{4}$.

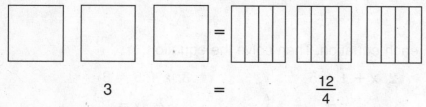

3 = $\frac{12}{4}$

1. Draw a model for 3. Then divide each whole into thirds.

   a. How many thirds are there? _____

   b. Use the model to write a fraction that is equal to 3. _____

2. Draw a model for 3. Then divide each whole into fifths.

   a. How many fifths are there? _____

   b. Use the model to write a fraction that is equal to 3. _____

3. What do you notice about all of the fractions that are equal to 3? How is the numerator related to the denominator?

   _____

4. Write two more fractions that are equal to 3.

   _____

Try This

Use models to help you write three fractions that are equal to each whole number.

1. 2 _____ **2.** 5 _____ **3.** 6 _____

Write the whole number that is equal to each fraction.

4. $\frac{18}{6}$ **5.** $\frac{20}{5}$ **6.** $\frac{8}{1}$

Holt Mathematics

TEKS 6.2.A

LAB 5-2A

Add Fractions with Fraction Circles

Use with Lesson 5-2

Materials needed: set of fraction circles
Activity
In this lab you will use fraction circles to help you add fractions with different denominators. Follow these steps to model $\frac{1}{2} + \frac{1}{3}$.

Objects	**Words**	**Numbers**

1.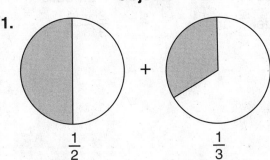

$\frac{1}{2}$ + $\frac{1}{3}$

Model $\frac{1}{2}$ and $\frac{1}{3}$.

$\frac{1}{2} + \frac{1}{3}$

2.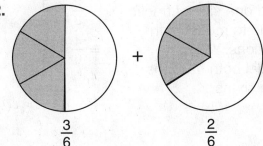

$\frac{3}{6}$ + $\frac{2}{6}$

Find one size of fraction circle to model both fractions. You can model both fractions with sixths.

$\frac{1}{2} + \frac{1}{3}$

$= \frac{3}{6} + \frac{2}{6}$

3.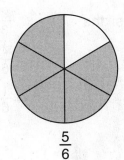

$\frac{5}{6}$

Combine the pieces to find out how many sixths there are altogether.

$\frac{1}{2} + \frac{1}{3}$

$= \frac{3}{6} + \frac{2}{6}$

$= \frac{5}{6}$

Try This

Use fraction circles to help you find each sum.

1. $\frac{1}{3} + \frac{1}{4} =$ _____

2. $\frac{1}{8} + \frac{1}{4} =$ _____

3. $\frac{3}{4} + \frac{1}{8} =$ _____

4. $\frac{1}{4} + \frac{2}{3} =$ _____

5. $\frac{1}{6} + \frac{3}{4} =$ _____

6. $\frac{3}{8} + \frac{1}{2} =$ _____

Holt Mathematics

TEKS 6.2.A

LAB 5-2B

Subtracting Fractions with Fraction Circles

Use with Lesson 5-2

Materials needed: set of fraction circles

Activity

In this lab you will use fraction circles to help you subtract fractions with different denominators. Follow these steps to model $\frac{2}{3} - \frac{1}{2}$.

Objects	Words	Numbers
1. 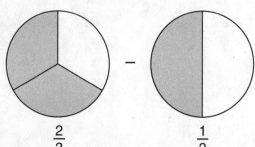 $\frac{2}{3}$ — $\frac{1}{2}$	Model $\frac{2}{3}$ and $\frac{1}{2}$.	$\frac{2}{3} - \frac{1}{2}$
2. 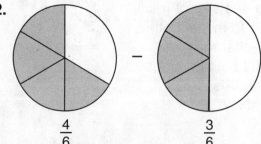 $\frac{4}{6}$ — $\frac{3}{6}$	Find one size of fraction circle to model both fractions. You can model both fractions with sixths.	$\frac{2}{3} - \frac{1}{2}$ $= \frac{4}{6} - \frac{3}{6}$
3. 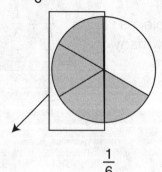 $\frac{1}{6}$	Take away 3 sixths from 4 sixths.	$\frac{2}{3} - \frac{1}{2}$ $= \frac{4}{6} - \frac{3}{6}$ $= \frac{1}{6}$

Try This

Use fraction circles to help you find each difference.

1. $\frac{5}{6} - \frac{1}{2} =$ _____

2. $\frac{7}{8} - \frac{3}{4} =$ _____

3. $\frac{1}{2} - \frac{1}{3} =$ _____

4. $\frac{11}{12} - \frac{1}{6} =$ _____

5. $\frac{5}{8} - \frac{1}{4} =$ _____

6. $\frac{3}{4} - \frac{2}{3} =$ _____

Holt Mathematics

TEKS 6.10.B

LAB 6-2A

Finding the Range, Mean, Median, and Mode

Use with Lesson 6-2

Activity

Find the range, mean, median, and mode of your classmates' shoe sizes.

1. Survey the class. List all shoe sizes below.

_____

_____

2. Find the mean. Find the sum of the numbers and then divide by the number of classmates.

_____

_____

3. Find the range. List the numbers in order from least to greatest. Subtract the smallest number from the largest one.

_____

_____

4. Find the median. Start with the list of numbers ordered from least to greatest. Find the number in the middle of this list or the mean of the two middle numbers.

_____

_____

5. Find the mode. Find the number that occurs the most often.

_____

_____

Think and Discuss

Suppose you run a factory that manufactures shoes. You need to decide what sizes of shoes to make and how many of each size to make.

1. How would the least and greastest data values you gathered be useful to you?

_____

_____

Holt Mathematics

TEKS 6.10.B

⭐ **LAB 6-2A** **Finding the Range, Mean, Median, and Mode**

2. What number would help you decide which shoe size would be the most popular to buy?

_____

_____

Try This

3 new students are added to your class.

1. Suppose that all of their shoe sizes are equal to the smallest size measured. How does the median change, if at all?

_____

_____

2. How does the mode change, if at all?

_____

_____

3. Now suppose that all of their shoe sizes are equal to the largest size measured. How does the median change, if at all?

_____

_____

4. How does the mode change, if at all?

_____

_____

5. How does the mean change, if at all?

_____

_____

Holt Mathematics

Name _____ Date _____ Class _____

TEKS 6.10.B

  Modeling the Mean, Median, Mode, and Range

Use with Lesson 6-2

Materials needed: pencil, centimeter cubes
Activity 1
You can think of the mean (or average) of a set of data as the point at which the values will balance. Imagine the number line as an infinitely long board. Place a weight on the number line for each value in the set of data. The mean is the balance point.

For the set of data consisting of two values, 6 and 12, the number line balances at 9. The mean of 6 and 12 is 9.

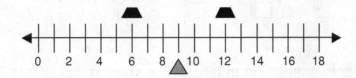

1. Seven students were asked how many hours they watched television over the weekend. Here are the results.

   3 6 6 9 2 3 6

   Draw a weight on the number line for each value in the set of data. You can stack the weights on a value if the value appears more than once in the set of data.

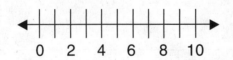

2. At what value do you think the number line would balance?

   _____

Holt Mathematics

Name _____ Date _____ Class _____

Modeling the Mean, Median, Mode, and Range

Activity 2

You can also use cubes to model the values in a set of data.

1. Make a stack of centimeter cubes to represent each value in the set of data from Activity 1.

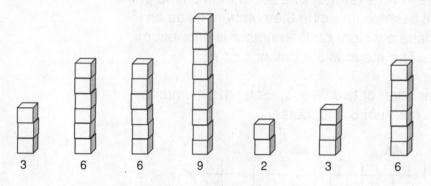

3   6   6   9   2   3   6

2. Line up the stacks in order from shortest to tallest. The stack in the middle of the row represents the median of the set of data. What is the median?

_____

3. The height of the stack that occurs most often is the mode of the set of data. What is the mode?

_____

4. Place the tallest stack next to the shortest stack. The difference in the heights of these stacks is the range of the set of data. What is the range?

_____

5. Rearrange the cubes so that all of the stacks have the same height. The new height of the stacks is the mean of the set of data. What is the mean?

_____

Try This

1. How does the mean that you found by rearranging the stacks compare to the mean that you found in Activity 1 by looking for a balance point?

_____

Holt Mathematics

Name _____ Date _____ Class _____

TEKS 6.10.B, 6.10.D

⭐ LAB 6-2C Finding Mean, Median, and Mode

Use with Lesson 6-2

Mean, median, and mode are all measures of a set of data. You can compute these measures with data from your class.

Activity

Find the mean, median, and mode of your classmates' heights in inches.

1. Survey the class. List all heights in inches below.

_____

_____

2. Find the sum of the heights and divide by the number of heights.

_____

_____

3. Then find the middle height or find the mean of the two middle heights.

_____

_____

4. Find the mode. Find the height that appears most often.

_____

_____

L17
Holt Mathematics

Name _____ Date _____ Class _____

TEKS 6.10.B, 6.10.D

⬟ **LAB**
6-2C **Finding Mean, Median, and Mode**

Think and Discuss

1. How do you compute the median for a set of data with an even number of numbers?

_____

_____

2. When can there be more than one mode in a set of data?

_____

_____

3. When will there be no mode in a set of data?

_____

_____

_____

Try This

1. Find your scores on your last five math tests. List them below.

_____

_____

2. What is the mean of these scores? _____

3. What is the median of these scores? _____

4. What is the mode of these scores? _____

5. What is the total number of pens and pencils you have with you in class? Ask three classmates for their totals as well. List the amounts below.

_____

_____

6. What is the mean of these numbers? _____

7. What is the median of these numbers? _____

8. What is the mode of these numbers? _____

Holt Mathematics

Name _____ Date _____ Class _____

TEKS 6.10.A, 6.10.D, 6.11.A

★ **LAB 6-5** # Market Research

Use with Lesson 6-5

Before a company begins to develop a new product, the company will usually conduct research to determine whether or not the product is likely to be successful. In this lab, you will collect information from your classmates to determine if a new ready-made lunch pack will be a hit.

The **population** you are interested in is all students in your school. However, you probably will not be able to ask everybody in your school about the new ready-made lunch pack. So you must choose a sample to ask. A **sample** is a group of individuals selected from a population. (Students in your school are a sample of all students.) You will also need to be sure that the students you do ask represent the opinions of all the students in your school. So you will need to use a random sample. **Random samples** are samples in which everyone in the population (students at your school) has an equal opportunity of being chosen.

Activity

1. Plan how you will choose your random sample. Write out your plan in the space below.

   _____

   _____

   _____

   _____

2. Design your new ready-made lunch pack. Try to stay close to some existing product, but with a new twist.

   _____

   _____

   _____

   _____

   _____

   _____

Holt Mathematics

TEKS 6.10.A, 6.10.D, 6.11.A

⭐ **LAB 6-5** **Market Research**

Use with Lesson 6-5

3. Design your survey and collect your information. Ask a question
   that students can answer quickly and easily. The answer must be
   numerical. In the space below, design your survey.

_____

_____

_____

_____

Think and Discuss

1. Examine your data. Can you determine if students would buy the
   ready-made lunch pack?

_____

_____

_____

2. Construct a bar graph of the responses. Construct a frequency
   table of the responses. Which data display helps you to better
   determine if students would buy your ready-made lunch pack?
   Explain.

_____

_____

Show your work.

Holt Mathematics

TEKS 6.10.A, 6.10.D, 6.11.A

⬟ LAB **Market Research**
6-5

Use with Lesson 6-5

3. Find the mean of the data. Does the mean help you to
determine if students would buy your ready-made lunch pack?
Explain.

_____

_____

_____

4. If your sample space was not a random selection of the
population, how might this affect your survey results? Provide an
example.

_____

_____

_____

_____

Try This

1. A class surveyed their classmates to determine
whether or not they liked deep-dish pizza. This was the
question that was asked.

Do you enjoy eating deep-dish pizza? Select a number
from 1 to 5 where 1 means that you do not enjoy eating
deep-dish pizza and 5 means you enjoy eating
deep-dish pizza very much.

Do not enjoy   Somewhat enjoy   Enjoy very much

1   2   3   4   5

The table shows the results of the survey.

a. Find the mean. _____

b. What does the mean tell you?

_____

_____

Student number	Response
1	3
2	5
3	5
4	2
5	3
6	2
7	5
8	4
9	2
10	4
11	4
12	5
13	1
14	3
15	1
Mean	

Holt Mathematics

Name _____ Date _____ Class _____

TEKS 6.10.A, 6.10.D, 6.11.A

★ LAB **6-5** **Market Research**

2. Write a new question that could be asked in a survey regarding deep-dish pizza. What random sample population would you use to conduct your survey?

_____

_____

_____

_____

_____

_____

3. Conduct a survey and make a table of your results. Find the mean and write one sentence explaining the results of your survey.

Student Number	Result	Student Number	Result

Mean _____   Results _____

Holt Mathematics

TEKS 6.12.B

LAB 6-10A

Evaluating Different Representations

Use with Lesson 6-10

Materials needed: pencil, graph paper or other materials for making graphs

Activity

Maria took a quiz in her science class every week for 12 weeks. The table shows Maria's scores on the quizzes.

Week	1	2	3	4	5	6	7	8	9	10	11	12
Score	72	75	81	80	81	79	80	87	87	91	93	95

1. Make a line plot of Maria's quiz scores.

2. Make a stem-and-leaf plot of Maria's quiz scores.

3. Make a line graph of Maria's quiz scores.

Holt Mathematics

Name _____ Date _____ Class _____

TEKS 6.12.B

Evaluating Different Representations

LAB 6-10A

Try This

1. Maria will get a B in the class if she can show that most of her quiz scores are in the 80s. Which of the three graphs should she choose to convince her teacher that she deserves a B? Explain your choice.

_____

_____

2. Maria's teacher says students will get extra credit if they can show that they improved over the 12-week period. Which of the three graphs should Maria choose to convince her teacher that she improved over the 12-week period? Explain your choice.

_____

_____

Holt Mathematics

TEKS 6.10.A

Choosing an Appropriate Display

Use with Lesson 6-10

Data can be represented in different ways, depending on the type of data and the message to be conveyed.

Type of Display	Best Use
Line graph	To show change in data over time
Bar graph	To show relationships or comparisons between groups
Circle graph	To compare parts to a whole
Line plot	To show distribution of a small data set in which the original data values are shown
Stem-and-leaf plot	To show distribution of a larger data set where frequency of values can be compared; original data values are shown

Activity

Average High Monthly Temperatures for Austin, Texas												
Month	Jan	Feb	Mar	Apr	May	Jun	Jul	Aug	Sep	Oct	Nov	Dec
Temperature (°F)	60	65	73	79	85	91	95	96	90	81	70	62

1. Which two displays represent the same data shown in the table?
(c and d are on the next page)

a.

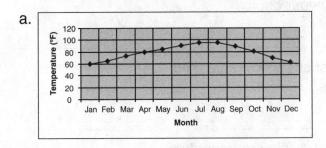

b.

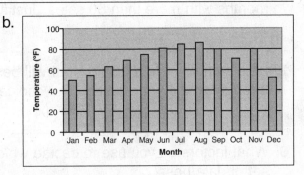

Holt Mathematics

TEKS 6.10.A

LAB
6-10B

Choosing an Appropriate Display

c.

Stems	Leaves
6	0 2 5
7	0 3 9
8	1 5
9	0 1 5 6

Key: 8|1 means 81

d.

```
                              X
                        X     X
              X     X   X     X
              X     X   X     X
             ─────────────────────
             60    70  80    90
```

2. Which display is best if you want to show how the temperature changes throughout the year? Explain.

_____

_____

3. Which display is best if you want to show that there are more months with temperatures in the 90's? Explain.

_____

_____

Try This

Use the data in the table to answer the following problems.

Number of Sales per Month at Ron's Refrigerators											
Month	Jan	Feb	Mar	Apr	May	Jun	Jul	Aug	Sep	Oct	Nov	Dec
Number of Sales	13	38	23	32	38	26	37	28	36	32	14	14

1. Select and create a display that will best show the number of months with more than 30 sales. Justify your choice.

_____

2. Select and create a display that will best show how sales did throughout the year. Justify your choice.

_____

3. What factors do you use to decide which display represents a set of data best?

_____

_____

Holt Mathematics

Name _____ Date _____ Class _____

Both the type of data and the message to be conveyed are important factors to consider when selecting a graphical display.

Activity

Marta works in the Marketing Communications department of a large company. It is her job to select and prepare graphical representations for company data. In each scenario below, help Marta decide how best to display the data. Justify your decision.

1. Marta has the company's average stock price per month for the last year.

Month	Jan	Feb	Mar	Apr	May	Jun	Jul	Aug	Sep	Oct	Nov	Dec
Stock Price ($)	18	19	19	21	25	23	28	30	31	30	32	33

a. The company wants to show stock holders how the stock price has risen during the last year.

_____

b. The company wants to show how many months the stock price was $30 or above a share.

_____

2. The company has three other major competitors. Marta has data for last month's sales for each company.

a. The company wants to compare the sales of all four companies.

_____

_____

b. The company wants to show how it led the industry last month in sales.

_____

Company	Number of Sales
Marta's Company	440
Competitor A	119
Competitor B	87
Competitor C	154

Holt Mathematics

Making Decisions with Data

3. Marta has the ages of all employees in the Finance department.

23   42   21   35   32   37   45   48   54   30

22   36   36   38   33   34   58   41   40   39

a. The Human Resources department wants to compare the ages of the department's employees. It wants to show how many employees are in their 20s, 30s, 40s, 50s, and 60s.

_____

_____

b. The department manager wants to know what percentage of the department's employees are between the ages of 20–29, 30–39, 40–49, and 50–59.

_____

4. Marta has the ages of the 15 children who attend the company's onsite daycare center.

3   3   4   4   3   3   4   5   4   4

3   4   5   4   4

a. The center's director wants a simple way to organize the data so that each original data value is shown.

_____

_____

b. For a parent presentation, a teacher wants to compare the totals for each age.

_____

_____

Try This

1. Explain why someone might use two different graphical representations of the same data.

_____

_____

_____

Holt Mathematics

Name _____ Date _____ Class _____

TEKS 6.10.D

LAB 6-10D **Collecting, Organizing, Displaying, and Interpreting Data**

Use with Lesson 6-10

Activity

Suppose your school is interested in providing new healthy snack choices. Design and conduct a survey to determine what types of snacks students would like. Organize your data and choose an appropriate display. Then interpret your data. Publish your findings. Suggest some new healthy snacks. Include supporting evidence for your suggestions.

Step 1: Design and conduct your survey.

a. Who are you going to survey (e.g., all students or a sample of students)?

_____

b. How will you survey your population (e.g., a written survey or a personal interview)?

_____

c. What questions will you ask? Be sure your questions are unbiased.

_____

d. When and where will you conduct your survey?

_____

Step 2: Organize and display your data.

a. How will you organize your data?

_____

b. What is the best display of your data (e.g., line plot, stem-and-leaf plot, bar graph, line graph, circle graph)?

_____

Step 3: Interpret your data.

a. What conclusions can you make?

_____

Holt Mathematics

TEKS 6.10.D

🟊 LAB 6-10D Collecting, Organizing, Displaying, and Interpreting Data

b. How does your data display support your conclusions?

_____

_____

Step 4: Publish your findings.

a. How will you publish your findings (e.g., article in the school paper, letter to the principal)?

_____

_____

b. How can you convince your audience that your data is reliable?

_____

_____

c. What evidence will you include to support your suggestions?

_____

_____

Try This

1. Conduct a survey about an issue in which you are interested. Follow the steps in the Activity.

_____

_____

2. Use two different data displays to represent your data. Explain when you might use each type of display.

_____

_____

_____

_____

Holt Mathematics

Name _____ Date _____ Class _____

TEKS 6.10.A

Displaying Basketball Data

Use with Lesson 6-10

Materials needed: pencil, graph paper or other materials for making graphs
Activity

The state of Texas has three teams in the National Basketball League (NBA). The table shows the number of wins for these teams over five seasons.

Number of Wins by NBA Teams from Texas					
Team	**2000–2001**	**2001–2002**	**2002–2003**	**2003–2004**	**2004–2005**
Dallas Mavericks	53	57	60	52	58
San Antonio Spurs	58	58	60	57	59
Houston Rockets	45	28	43	45	51

For each situation, tell whether you would use a line plot, line graph, bar graph, or stem-and-leaf plot to display the data. Then make the graph.

1. You want to display the number of wins by the three teams in the 2004–2005 season.

_____

Holt Mathematics

TEKS 6.10.A

Displaying Basketball Data

2. You want to show how the number of wins by the Houston Rockets changed over the five seasons.

_____

3. You want to display the number of wins by all the teams over all five seasons. For this situation, which two types of graphs could you use?

_____

Holt Mathematics

TEKS 6.10.A

⭐ 6-10E Displaying Basketball Data

Try This

1. Which of the graphs that you made let you compare teams?

_____

2. Which of the graphs that you made show the frequency of data values?

_____

_____

_____

Holt Mathematics

Name _____ Date _____ Class _____

TEKS 6.10.A

Representing Data in Graphs

LAB 6-10F

Use with Lesson 6-10

Activity

Use the data in the table below to answer the questions that follow.

Annual Number of Sunny Days for Major Western US Cities

City	Number of days of sunshine
Phoenix	211
Seattle	58
Los Angeles	186
San Francisco	160
Portland	68
Las Vegas	210
San Diego	146
Denver	115
Salt Lake City	125

Which of the following types of graphs would be appropriate to display the data? For each type of graph, either write NA for "Not Appropriate," or say what aspect of the data would be shown by using such a graph.

1. Line graph _____

2. Bar graph _____

3. Circle graph _____

4. Line plot _____

5. Stem-and-leaf plot _____

Think and Discuss

1. If you were going to display the data as a bar graph, would a horizontal bar graph be more or less convenient than a vertical bar graph? Why?

_____

_____

Holt Mathematics

TEKS 6.10.A

Representing Data in Graphs

2. Think of two ways you could order the data in a bar graph, and what advantage there could be in each.

_____

_____

_____

3. Use the maximum and minimum values in the table to find an appropriate scale and interval for a bar graph based on the data.

   a. Scale _____   **b.** Interval _____

4. Fill in the bar graph below. Be sure to label the scale and interval.

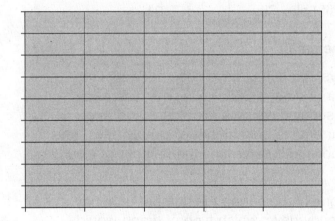

5. Use the graph you created to answer the following questions.

   a. Which two cities had the greatest number of sunny days?

_____

   b. Which city had the least number of sunny days?

_____

Holt Mathematics

Name _____ Date _____ Class _____

Representing Data in Graphs

Try This

The table below shows the number of days of rain for the same western US cities.

Annual Number of Rainy Days for Major Western US Cities

City	Number of days of sunshine
Phoenix	36
Seattle	151
Los Angeles	35
San Francisco	68
Portland	153
Las Vegas	26
San Diego	41
Denver	89
Salt Lake City	91

1. Create a bar graph of the data.

2. The number of rainy days and sunny days does not equal the number of days in a year (1 year = 365 days). The difference between the number of sunny days plus rainy days and the number of days in a year equals the number of cloudy days. Add a column to the table above and determine the number of cloudy days for each city.

3. If you were to create a graph comparing the number of sunny days, rainy days, and cloudy days, which type of graph would you use? Explain.

_____

_____

_____

Holt Mathematics

Name _____ Date _____ Class _____

 LAB 7-1A

Represent Ratios with Decimals

Use with Lesson 7-1

Materials needed: pencil, decimal grids or graph paper
Activity

You can use decimal grids to represent ratios with decimals.

A 10-by-10 grid represents 1. Each column or row of the grid represents one tenth or 0.1. Each small square represents one one-hundredth or 0.01.

Follow these steps to represent the ratio 3 to 5 (or $\frac{3}{5}$) with a decimal.

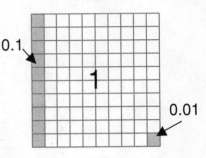

0.1

1

0.01

1. Divide the decimal grid into 5 equal parts.

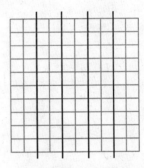

2. Shade 3 of the 5 parts.

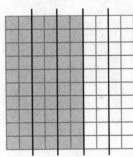

3. Count the number of squares that are shaded. Six columns are shaded and each column represents 0.1, so 6 columns represent 0.6. $\frac{3}{5} = 0.6$

Holt Mathematics

Name _____ Date _____ Class _____

TEKS 6.3.B.03

Represent Ratios with Decimals

Try This

Use decimal grids to represent each ratio as a decimal.

1. 3 to 4 _____

2. 7 to 10 _____

3. $\frac{1}{4}$ _____

4. $\frac{3}{20}$ _____

Holt Mathematics

Name _____ Date _____ Class _____

TEKS 6.08.A

Volume

Use with Lesson 7-1

Activity

Find the capacity of this container.

To find the capacity, look at the top of the fluid in the container. Then, read the measurement on the measuring cup.

$1\frac{3}{4}$ cups

Think and Discuss

1. Examine the measuring devices you have in your classroom (measuring cup, beaker, graduated cylinder). Why is a graduated cylinder a better tool to measure milliliter capacity than a measuring cup? Explain.

_____

_____

Try This

1. Estimate how many cups of water will fit into one liter. Then using your measuring cup, fill a one-liter beaker with water. Keep track of how many cups it actually takes to fill the beaker. Compare your actual results with your estimate.

_____

2. Estimate how many milliliters of water are needed to fill a 2-cup container. Then using your graduated cylinder, fill a 2-cup container with water. Keep track of how many milliliters it actually takes to fill the container. Compare your actual results with your estimate.

_____

3. Fill an empty water bottle with sand. Then pour the sand from the bottle into an empty 2-liter bottle. Approximately how many water bottles of sand does it take to fill one 2-liter bottle? Estimate how many water bottles of sand it would take to fill 100 2-liter bottles.

_____

Holt Mathematics

Name _____ Date _____ Class _____

TEKS 6.08.A

  Temperature

Use with Lesson 7-1

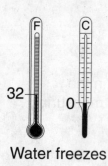

Water freezes

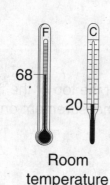

Room
temperature

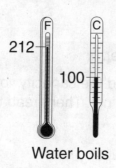

Water boils

Activity

Estimate the temperature of the snowman in °F and °C.

A reasonable temperature for making a snowman is 28°F or
−2°C, since that is below the point when water freezes.

Use the thermometer to find the temperature of this pot of
boiling soup, in °F and °C.

The thermometer reads 212°F and 100°C.

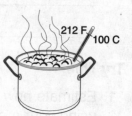

Think and Discuss

1. Which temperature is a teapot boiling at, 100°F or 100°C?
   Explain how you know.

   _____

   _____

Try This

Use your thermometer and measure the temperature of each of the
following.

1. a cup of water taken from a drinking fountain _____

2. a cup of cold water taken from a bathroom sink _____

3. a cup of warm water taken from a bathroom sink _____

4. a beverage, such as a juice box _____

5. a melted ice cube _____

Holt Mathematics

Name _____ Date _____ Class _____

TEKS 6.6.3

Modeling Percents and Ratios

LAB 7-7

Use with Lesson 7-7

Materials needed: sheet of paper, ruler, scissors, pencil

Activity

In this lab you will make a strip of paper to model percents and a strip of paper to model ratios. You can use the two strips together to write percents as ratios and vice versa.

1. Cut two 10-inch strips from a sheet of paper. The strips should each be about an inch wide.

2. On one strip, use the ruler to make marks at 1 in., 2 in., 3 in., and so on up to 9 in.

3. Label the mark at 1 in. as 10%. Label the mark at 2 in. as 20%. Continue in this way up to 90%.

4. Make marks at $2\frac{1}{2}$ in. and at $7\frac{1}{2}$ in. and label these as 25% and 75%.

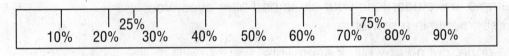

5. On the second strip, make marks at 2 in., 4 in., 6 in., and 8 in. Then label these as $\frac{1}{5}$, $\frac{2}{5}$, $\frac{3}{5}$, and $\frac{4}{5}$.

6. Finally, make marks at $2\frac{1}{2}$ in., 5 in., and $7\frac{1}{2}$ in. and label these as $\frac{1}{4}$, $\frac{1}{2}$, and $\frac{3}{4}$.

$$\frac{1}{5} \quad \frac{1}{4} \qquad \frac{2}{5} \quad \frac{1}{2} \qquad \frac{3}{5} \qquad \frac{3}{4} \quad \frac{4}{5}$$

7. Line up the two strips to see equivalent percents and ratios.

Try This

Use the strips to write each percent as a ratio and each ratio as a percent.

1. 20% _____   2. $\frac{3}{5}$ _____   3. 80% _____   4. $\frac{3}{4}$ _____

Holt Mathematics

TEKS 6.10.C

Making a Circle Graph to Display Data

Use with Lesson 7-10

Materials needed: scissors; brown, green, and blue markers or crayons; tape; blank sheet of paper

Activity

Survey your class to determine everyone's eye color. Use a table to collect your data.

Eye Color	Tally	Total Number
Brown		
Green or Hazel		
Blue or Gray		

Now, cut out one bar found on the last page of this lab. Each segment of the bar represents one student who participated in your survey.

Color the bar to show the number of students who had each eye color. For example, if 6 students in your class had blue eyes, you would color 6 segments in a row blue.

When you are done, cut off any blank segments.

Join the ends of your bar to form a round loop with the colored side facing inwards. Tape the ends together.

Place the loop on a blank piece of paper and trace around the inside edge to draw a circle. Mark where the colors meet. Remove the loop.

Make a dot at the center of your circle. Connect the marks at the edge of your circle with the center dot to form sections. Color each section to match the related color on the inside of the loop. You have just used a segmented bar chart to make a **circle graph** of your data!

1. How are the segmented bar chart and circle graph related?

_____

_____

_____

_____

Holt Mathematics

TEKS 6.10.C

LAB
7-10

Making a Circle Graph to Display Data

2. What different information does each data display provide?

_____

_____

_____

3. How many total students participated in your survey? _____

4. What fraction of the students surveyed have each eye color?

brown _____   green _____   blue _____

_____

5. What percent of students surveyed have each eye color? Make an estimate if needed.

brown _____   green _____   blue _____

_____

6. What does the entire circle graph represent? Explain.

_____

_____

_____

7. Finish your circle graph by writing a title, making a key, and labeling each section.

8. Write two conclusions you can make from your circle graph.

_____

Try This

1. Conduct a survey about a topic in which you are interested. Collect your data. Then organize and display your data.

_____

2. How did you organize your data?

_____

3. How did you decide the best representation of your data?

_____

Holt Mathematics

Making a Circle Graph to Display Data

LAB 7-10

4. Could you represent your data a different way? Explain.

_____

5. Write two conclusions you can draw from your data.

_____

_____

_____

_____

_____

_____

Holt Mathematics

TEKS 6.10.C

Making a Circle Graph to Display Data

<table>
<tr><td></td><td></td></tr>
<tr><td></td><td></td></tr>
<tr><td></td><td></td></tr>
<tr><td></td><td></td></tr>
<tr><td></td><td></td></tr>
<tr><td></td><td></td></tr>
<tr><td></td><td></td></tr>
<tr><td></td><td></td></tr>
<tr><td></td><td></td></tr>
<tr><td></td><td></td></tr>
<tr><td></td><td></td></tr>
<tr><td></td><td></td></tr>
<tr><td></td><td></td></tr>
<tr><td></td><td></td></tr>
<tr><td></td><td></td></tr>
<tr><td></td><td></td></tr>
<tr><td></td><td></td></tr>
<tr><td></td><td></td></tr>
<tr><td></td><td></td></tr>
<tr><td></td><td></td></tr>
<tr><td></td><td></td></tr>
<tr><td></td><td></td></tr>
<tr><td></td><td></td></tr>
</table>

Holt Mathematics

Name _____ Date _____ Class _____

TEKS 6.10.C

 LAB 7-10B **Choose a Circle Graph that Best Displays Data**

Use with Lesson 7-10

Activity

Circle graphs show parts of a whole. The entire circle represents 100% of the data. Each section of the circle graph represents a percent of the total. Select the circle graph that best displays the data in the table.

A survey asked 400 students to name their favorite color. The results are shown in the table.

Color	Number of Students
Blue	72
Green	68
Purple	40
Red	124
Yellow	96

A

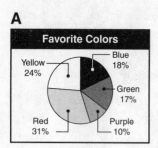

B

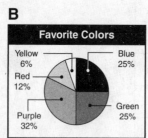

C

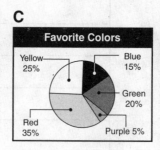

D

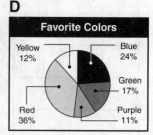

1. What percent of the students named each color as his or her favorite?

_____

2. Which graph best displays the results of the survey? Explain.

_____

Try This

1. Which data set can be displayed by the circle graph? Explain.

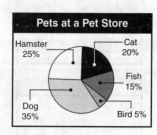

_____

A

Pet	Number
Cat	72
Fish	5
Bird	12
Dog	60
Hamster	15

B

Pet	Number
Cat	36
Fish	27
Bird	9
Dog	63
Hamster	45

C

Pet	Number
Cat	40
Fish	30
Bird	5
Dog	70
Hamster	25

Holt Mathematics

Name _____ Date _____ Class _____

Classifying Angles

Use with Lesson 8-2

Materials needed: protractor
Activity
Find the measure of each angle. Then make a conjecture to define each type of angle.

1.

Group A: Acute Angles	Not Acute Angles

Definition of acute angle: _____

2.

Group B: Right Angles	Not Right Angles

Definition of right angle: _____

Holt Mathematics

Name _____ Date _____ Class _____

LAB 8-2

Classifying Angles

3.

Group C: Obtuse Angles	Not Obtuse Angles

Definition of obtuse angle: _____

_____

Try This

Measure 8 angles found in objects around you. Classify the angles as acute, right, or obtuse. Try to find each type of angle at least once.

	Object Measured	Angle Measure	Angle Name
1.			
2.			
3.			
4.			
5.			
6.			
7.			
8.			

Holt Mathematics

TEKS 6.6.A

Classifying Angles

LAB
8-2

9. Is it easier to find certain types of angles more than others? Explain.

_____

_____

10. Do certain shapes have certain types of angles? Explain.

_____

_____

_____

_____

_____

_____

Holt Mathematics

Name _____ Date _____ Class _____

 LAB 8-10 ## Mathematics in Music

Use with Lesson 8-10

Twinkle Twinkle Little Star

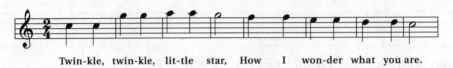

Notes that look like this ♩ are called quarter notes. Notes that look like this ♩ are called half notes. Quarter notes get one beat. Half notes get two beats. A measure is the portion of the staff between two vertical lines. This is one measure: [▦]. Most music has a time signature that indicates how many beats are in a measure and which note gets the beat. The time signature for "Twinkle Twinkle Little Star" is $\frac{2}{4}$, indicating there are 2 beats per measure and a quarter note gets one beat. Other common time signatures are $\frac{3}{4}$ and $\frac{4}{4}$. In $\frac{3}{4}$ time (waltz time), there are three beats per measure.

Activity

1. Sing or hum the first line to the song. Tap a pencil on the table to keep time. How many beats are in the first line?

   _____

2. There are two phrases in the line. The two words that rhyme end the phrases. How many beats are in each phrase?

   _____

Holt Mathematics

TEKS 6.11.A

⭐ 🔬 LAB 8-10 Mathematics in Music

Think and Discuss

1. A famous hymn is written in $\frac{3}{2}$ time. How many beats are in one measure?

   _____

2. A folk dance is written in $\frac{7}{4}$ time. What note gets one beat?

   _____

Try This

1. A famous jazz composition by Dave Brubek is called "Take Five." Its time signature is $\frac{5}{4}$. How many beats are in each measure?

   _____

2. In "Take Five," what note gets one beat?

   _____

Holt Mathematics

Name _____ Date _____ Class _____

TEKS 6.8.B.06

Solving Problems Involving Time

Use with Lesson 9-5

Materials needed: pencil
Activity
You can sketch the face of a clock to help you solve problems involving time.

An airplane takes off at 9:50 A.M. The flight lasts 6 hours 20 minutes. At what time does the plane land?

1. Sketch the face of a clock. Draw hands to show the starting time of 9:50.

2. Start at the 9 and add 6 hours. The resulting time is 3:50.

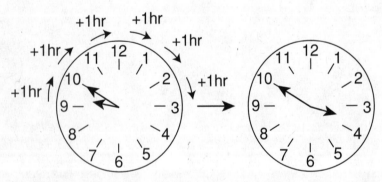

3. Now add 20 minutes. The resulting time is 4:10, so the plane lands at 4:10 P.M.

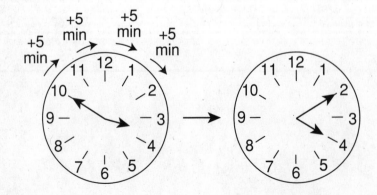

Holt Mathematics

TEKS 6.8.B.06

Solving Problems Involving Time

LAB 9-5

Try This

1. Miguel starts riding his bike at 11:40 A.M. He rides for 2 hours 30 minutes. At what time does he stop riding?

   _____

2. A film begins at 7:15 P.M. The film lasts 2 hours 50 minutes. At what time does the film end?

   _____

Holt Mathematics

Name _____ Date _____ Class _____

TEKS 6.08.A

Perimeter

LAB 9-7

Use with Lesson 9-7

The perimeter of an object is the distance around the object. Being able to find the perimeter of a polygon is handy for anyone who wants to build a fence, plant a border for a garden, buy gutters for a house, and many other applications.

Activity

1. Three objects are shown below: a stop sign, a yield sign, and a school crossing sign. The stop sign is an octagon, the yield sign is a triangle, and the school crossing sign is a pentagon.

2. Estimate the perimeter of each object in centimeters. Then determine the perimeter of each object by measuring the length of each side and adding the lengths together. Complete the following table.

Sign	Number of sides	Estimated Perimeter	Actual Perimeter
Stop sign			
Yield sign			
School crossing sign			

3. Now estimate the perimeter of your desktop. Then measure the actual length of each side. Add the lengths to find the perimeter of your desktop. Are any of the sides equal to any of the other sides?

_____

_____

Holt Mathematics

TEKS 6.08.A

  Perimeter

LAB
9-7

4. You can use a shortcut to determine the perimeter of a rectangle. The opposite sides of a rectangle are equal, so you can measure any two sides that touch, add their lengths, and then double the answer. Estimate the perimeter of this desktop.

60 cm

40 cm

Think and Discuss

1. If a polygon is a regular polygon, you can determine the perimeter by measuring only one side and then multiplying by the number of sides. This works for regular polygons because all of the sides are the same length. Which of the signs from the Activity are regular polygons?

_____

2. Would this procedure work for the school crossing sign?

_____

Try This

1. A rectangle has a length of 8 cm and a width of 3 cm. First, estimate the perimeter. Then calculate the actual perimeter.

3 cm

8 cm

_____

2. One side of a regular pentagon is 3 in. What is its perimeter?

3 in.

_____

3. A gardener is designing a flowerbed. The bed will have the shape shown. Estimate how many meters of edging will be necessary to surround the bed.

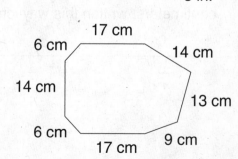

17 cm
6 cm   14 cm
14 cm
13 cm
6 cm
9 cm
17 cm

_____

Holt Mathematics

Name _____ Date _____ Class _____

TEKS 6.08.A, 6.08.B

Estimating and Finding Area

LAB 10-1

Use with Lesson 10-1

The area of a figure is the amount of surface it covers. Being able to measure area is important to engineers, gardeners, painters, architects, and many others. The area of many figures can be calculated from the figure's dimensions. For irregular or oddly shaped figures, the area can be estimated.

Activity

10 cm

4 cm

1. The rectangle above is 10 cm wide and 4 cm high. Cut several squares of paper 1 cm by 1 cm. Lay the squares of paper out on the rectangle and determine the number of squares it would take to cover the rectangle.

_____

2. Multiply the width by the height of the rectangle. Compare the product with the result you got in the first step. How do they compare?

_____

3. For all rectangles, the area can be calculated by multiplying the length by the width. The formula is $A = \ell \times w$. Use this formula to measure the area of your desktop. The result will be in square centimeters, written this way: cm^2.

_____

Holt Mathematics

Estimating and Finding Area

LAB 10-1

4. There are simple methods for calculating the area of many other geometric figures. Copy this parallelogram onto a separate sheet of paper. Using scissors, carefully cut the small triangle from the left side of the parallelogram and move it to the right side as shown. What new geometric figure is formed? _____

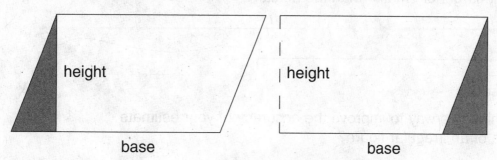

What formula can you use to find the area of this new figure? _____

How does the area of the new figure compare to the original area? _____

Calculate the area of a parallelogram with a height of 5 m and a base of 11 m. _____

5. Copy this triangle onto a sheet of paper two times. Cut out one of the triangles and turn it over. Can you put the two triangles together to form a parallelogram? _____

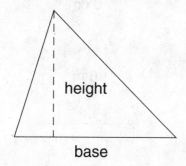

What formula can you use to find the area of the parallelogram formed by putting the two triangles together? _____

How does the area of the parallelogram formed by the two triangles compare to the area of the original triangle? _____

This shows that area of a triangle is $A = \frac{1}{2} \cdot b \cdot h$, where b is the base and h is the height. Calculate the area of a triangle with a base of 6 in. and a height of 3 inches. _____

Holt Mathematics

LAB
10-1   # Estimating and Finding Area

6. The area of irregularly shaped areas can be estimated. Trace the figure below onto centimeter grid paper. Count the number of complete squares within the figure.

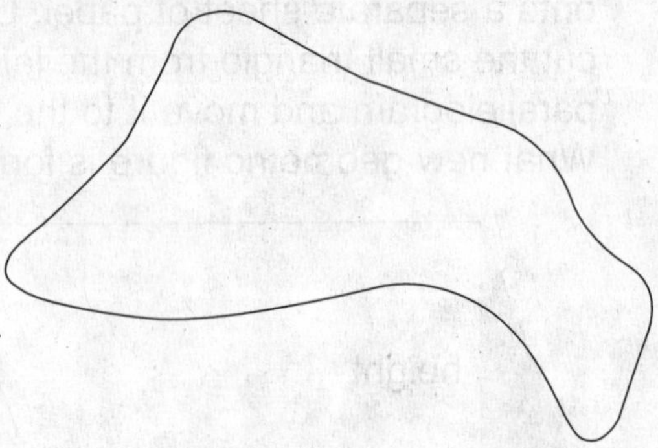

_____

Is this value larger or smaller than the actual area?

_____

Think and Discuss

1. Can you think of a way to improve the accuracy of your estimate of the area of an irregular figure?

_____

_____

_____

Try This

1. A rectangle has a length of 8 cm and a width of 3 cm. Find its area.

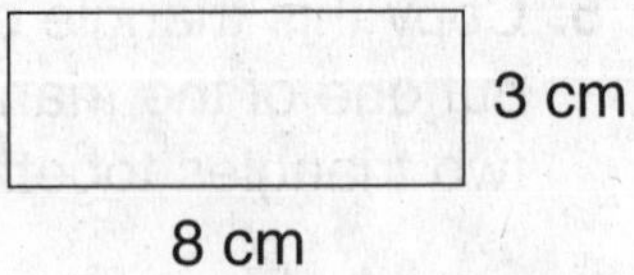

_____

3 cm

8 cm

2. A triangle has the dimensions shown. What is its area?

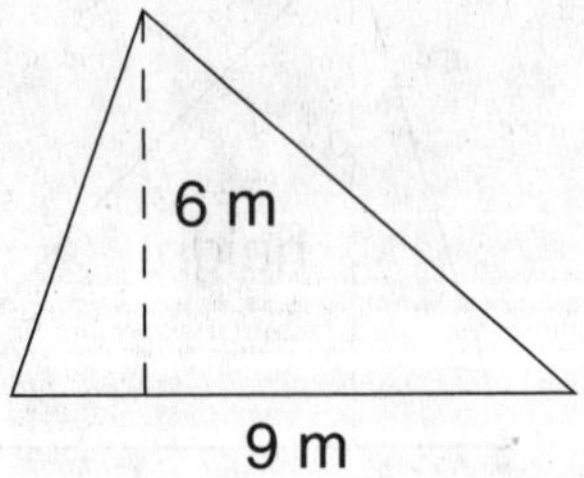

_____

6 m

9 m

Holt Mathematics

Name _____ Date _____ Class _____

TEKS 6.4.B

Use Tables of Data to Generate Formulas

LAB
10-7A

Use with Lesson 10-7

Materials needed: pencil
Activity
In this lab you will look for patterns in tables of data to discover
some formulas for rectangular prisms.

The rectangular prism at right has length ℓ, width *w*, and
height *h*.

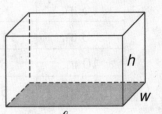

1. Look at the base of the prism. The base is the rectangular
    face that is shaded in the figure. The table shows different
    lengths and widths for the base and the perimeter of
    the base.

Length, ℓ	Width, *w*	Perimeter, *P*
5 in.	4 in.	18 in.
3 in.	1 in.	8 in.
10 in.	6 in.	32 in.
7 in.	5 in.	24 in.

a. Look for a pattern in the table. How is the perimeter related to
    the length and width? (*Hint:* First add the length and width.)

    _____

b. Suppose the length of the base is 9 in. and the width is 8 in.
    How can you find the perimeter?

    _____

c. Suppose the length of the base is ℓ and the width is *w*. How
    can you find the perimeter? Write your answer by completing
    the formula *P* = _____

Holt Mathematics

LAB 10-7A ## Use Tables of Data to Generate Formulas

2. The table shows different lengths and widths for the base and the area of the base.

Length, ℓ	Width, w	Area, A
5 in.	4 in.	20 in.^{2}
3 in.	1 in.	3 in.^{2}
10 in.	6 in.	60 in.^{2}
7 in.	5 in.	35 in.^{2}

a. Look for a pattern in the table. How is the area related to the length and width?

_____

b. Suppose the length of the base is ℓ and the width is w. How can you find the area? Write your answer by completing the formula $A =$ ____

3. The table shows different lengths, widths, and heights for the prism and the volume of the prism.

Length, ℓ	Width, w	Height, h	Volume, V
5 in.	4 in.	2 in.	40 in.^{3}
3 in.	1 in.	4 in.	12 in.^{3}
10 in.	6 in.	3 in.	180 in.^{3}
7 in.	5 in.	2 in.	70 in.^{3}

a. Look for a pattern in the table. How is the volume related to the length, width, and height?

_____

b. Suppose the length of the prism is ℓ, the width is w, and the height is h. How can you find the volume? Write your answer by completing the formula $V =$ ____

Holt Mathematics

TEKS 6.4.B

Use Tables of Data to Generate Formulas

Try This

Use your formulas to find the following.

1. The perimeter of the base of the prism

_____

2. The area of the base of the prism

_____

3. The volume of the prism

_____

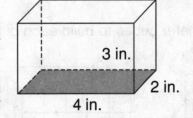

3 in.

2 in.

4 in.

Holt Mathematics

Name _____ Date _____ Class _____

Materials needed: centimeter cubes, pencil
Activity

1. Use centimeter cubes to build each of the prisms shown below.

Prism A

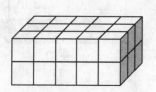
Prism B

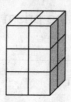
Prism C

2. a. For each prism, find the length, width, and perimeter of the base. Record the data in the table.

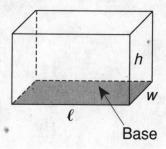

	Length, ℓ	Width, w	Perimeter, P
Prism A			
Prism B			
Prism C			

b. How is the perimeter related to the length and width? (*Hint:* First add the length and width.)

_____

c. Suppose the base of a prism has length ℓ and width w. Complete the formula for the perimeter, P.
$P =$ _____

Holt Mathematics

Name _____ Date _____ Class _____

LAB 10-7B

Completing Tables to Generate Formulas

3. **a.** For each prism, find the length, width, and area of the base.
Record the data in the table.

	Length, ℓ	Width, w	Area, A
Prism A			
Prism B			
Prism C			

b. How is the area related to the length and width?

_____

c. Suppose the base of a prism has length ℓ and
width w. Complete the formula for the area, A.
$A =$ _____

4. **a.** For each prism, find the length, width, and height. Also count
the number of cubes it takes to build the prism. The number
of cubes is the prism's volume. Record the data in the table.

	Length, ℓ	Width, w	Height, h	Volume, V
Prism A				
Prism B				
Prism C				

b. How is the volume related to the length, width, and height?

_____

c. Suppose a prism has length ℓ, width w, and height h.
Complete the formula for the volume, V. $V =$ _____

Holt Mathematics

LAB 10-7B Completing Tables to Generate Formulas

Try This

Use your formulas to find the following.

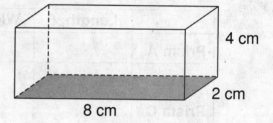

4 cm

2 cm

8 cm

1. The perimeter of the base of the prism

   _____

2. The area of the base of the prism

   _____

3. The volume of the prism

   _____

Holt Mathematics

TEKS 6.8.B

Select Appropriate Units of Measure

Use with Lesson 10-7

Materials needed: pencil
Activity
The Heard Natural Science Museum in McKinney, Texas, features an exhibit of poisonous snakes.

Complete each of the following statements about the exhibit by choosing the appropriate units from the tables. A benchmark is given for each unit to help you make your choice.

1. The exhibit includes a western diamondback rattlesnake. This snake can grow to a length of 5 _____.

Unit	Benchmark
Inch	Width of your thumb
Foot	Distance from your shoulder to your elbow
Mile	Total length of 18 football fields

2. The exhibit's rattlesnakes live in a rectangular enclosure. The perimeter of the enclosure is approximately 10 _____.

Unit	Benchmark
Centimeter	Width of a fingernail
Meter	Width of a single bed
Kilometer	Distance around a city block

3. The rattlesnake enclosure has an area of 60 _____.

Unit	Benchmark
Square inch	Area of a postage stamp
Square foot	Area of the cover of a textbook
Square mile	Area of approximately 250 city blocks

Holt Mathematics

Name _____ Date _____ Class _____

TEKS 6.8.B

LAB
10-7C

Select Appropriate Units of Measure

4. A single bite from a western diamondback rattlesnake can deliver as much as 4 _____ of venom.

Unit	Benchmark
Milliliter	Volume of a drop of water
Liter	Volume of a blender container
Kiloliter	Volume of 2 large refrigerators

5. Rattlesnakes are often found in the desert. They prefer temperatures of approximately 25 _____.

Unit	Benchmark
Degrees Fahrenheit	Water freezes at 32° Fahrenheit Water boils at 212° Fahrenheit
Degrees Celsius	Water freezes at 0° Celsius Water boils at 100° Celsius

6. The largest western diamondback rattlesnakes can weigh as much as 14 _____.

Unit	Benchmark
Ounce	Weight of a slice of bread
Pound	Weight of a loaf of bread
Ton	Weight of a small car

7. On Sundays, the Heard Natural Science Museum is only open in the afternoon. It is open for a total of 240 _____.

Unit	Benchmark
Second	The time between heartbeats
Minute	The length of two TV commercials
Hour	The length of two TV comedy programs

Holt Mathematics

Name _____ Date _____ Class _____

TEKS 6.8.B

LAB
10-7C
Select Appropriate Units of Measure

Try This

Select an appropriate unit that could be used to measure each of
the following. There may be more than one correct response.

1. The length of a rattlesnake's rattle _____

2. The area of a viewing window at a museum _____

3. The weight of a rattlesnake egg _____

4. The temperature of the water in an aquarium

_____

5. The volume of a home aquarium _____

Holt Mathematics

Name _____ Date _____ Class _____

TEKS 6.8.B

LAB 10-7D

Select Measurement Tools

Use with Lesson 10-7

Materials needed: pencil

Activity

For each situation, choose the appropriate measurement tool from the list below.

measuring tape marked in inches	meter stick
scale	measuring cup
thermometer	stopwatch

1. Latrell wants to know the perimeter of a painting in centimeters.

   _____

2. Jemma is following a recipe. She needs to check that a bag of potatoes weighs 2 pounds.

   _____

3. Brian wants to know how long it takes him to run one mile.

   _____

4. Mei wants to know the area of a tabletop in square inches.

   _____

5. Carlos wants to add 2 fluid ounces of bleach to the water in his washing machine.

   _____

6. Anne wants to check that her refrigerator is keeping food at 40°F.

   _____

Try This

1. Name all of the tools from the list that could be used to find the volume of a small aquarium in the shape of a rectangular prism.

   _____

Holt Mathematics

TEKS 6.8.B

Select Measurement Formulas

LAB 10-7E

Use with Lesson 10-7

Materials needed: pencil
Activity
For each situation, select an appropriate formula from the list.

Measurement Formulas		
Perimeter of a Rectangle	$P = 2\ell + 2w$	ℓ = length, w = width
Area of a Rectangle	$A = \ell w$	ℓ = length, w = width
Area of a Triangle	$A = \frac{1}{2}bh$	b = base, h = height
Volume of a Rectangular Prism	$V = \ell wh$	ℓ = length, w = width, h = height
Temperature Conversions	$C = \frac{5}{9}(F - 32)$ $F = \frac{9}{5}C + 32$	C = temperature in degrees Celsius F = temperature in degrees Fahrenheit
Weight Conversions	$Z = 16P$ $P = 2000T$	Z = weight in ounces P = weight in pounds T = weight in tons
Time Conversions	$S = 60M$ $M = 60H$	S = time in seconds M = time in minutes H = time in hours

1. Ryan uses a stopwatch to find out how many minutes he runs.
   He wants to know how many seconds this is. _____

2. The low temperature for January 10 is given in degrees Celsius.
   Susan wants to know how many degrees Fahrenheit this is.

   _____

3. Donnell measures the dimensions of a rectangular plot in his
   garden. He wants to know the distance around the plot.

   _____

4. Jordan has a cardboard box in the shape of a rectangular prism.
   He knows the lengths of all the edges of the box and he wants
   to calculate the box's volume. _____

5. At the zoo, the weight of a hippopotamus is given in tons. Alison
   wants to know how many pounds this is. _____

Holt Mathematics

Select Measurement Formulas

6. Flora takes her temperature using a thermometer marked in degrees Fahrenheit. She wants to know her temperature in degrees Celsius. _____

7. Kenji looks up the length and width of the *Mona Lisa*. He wants to know the area of the painting. _____

8. A recipe calls for 36 ounces of tomatoes. Jennifer has a can of tomatoes that gives the weight in pounds. She wants to know if she has the correct number of ounces. _____

9. Mike cuts out a triangular piece of cloth for a quilt. He measures the cloth and needs to know if its area is greater than 25 cm². _____

10. The running time of a DVD is given in hours. Wei wants to know if the DVD is longer than 150 minutes. _____

Try This

Describe a situation in which each formula might be used.

1. $V = \ell wh$

_____

_____

2. $P = 2\ell + 2w$

_____

_____

3. $C = \frac{5}{9}(F - 32)$

_____

_____

4. $M = 60H$

_____

_____

Holt Mathematics

Select Measurement Formulas

5. $A = \ell w$

_____

_____

6. $F = \frac{9}{5}C + 32$

_____

_____

Holt Mathematics

TEKS 6.8.B

Use Appropriate Units of Measure

LAB 10-7F

Use with Lesson 10-7

Materials needed: meter stick or tape measure, stopwatch, scale, thermometer
Activity
In this lab, you will work with classmates to make estimates and take measurements in your classroom.

1. Work in a small group. As a group, decide on an estimate for the length of your classroom. Record the estimate below, using appropriate units.

   _____

2. Now use a meter stick or tape measure to measure the length of the classroom. Be sure to use the same units you used in your estimate. Record the length below.

   _____

3. How does the actual length of the classroom compare to the group's estimate?

   _____

4. As a group, decide on an estimate for the width of the classroom.

   _____

5. Measure the width of the classroom.

   _____

6. As a group, decide on an estimate for the height of the room. Since the height can be difficult to measure directly, you will use this estimate as the actual height.

   _____

7. Calculate the perimeter of the classroom's floor. Be sure to use the actual measurements and appropriate units.

   _____

8. Calculate the area of the classroom's floor. Be sure to use appropriate units.

   _____

Holt Mathematics

TEKS 6.8.B

✪ **LAB 10-7F** **Use Appropriate Units of Measure**

9. Calculate the volume of the classroom. Be sure to use appropriate units.

_____

10. As a group, estimate the length of time it will take a student to walk across the room at a normal pace. Be sure to use appropriate units.

_____

11. Use a stopwatch to time a student as he or she walks across the room at a normal pace.

_____

12. As a group, decide on an estimate for the temperature of the room. Be sure to use appropriate units.

_____

13. Use a thermometer to find the actual temperature of the room.

_____

14. Choose a small object in the classroom and decide on an estimate for the object's weight. Be sure to use appropriate units.

_____

15. Weigh the object on a scale to find the actual weight.

_____

Try This

Describe an object in the classroom that can be measured with the given units.

1. Ounces

_____

2. Centimeters

_____

3. Square meters

_____

Holt Mathematics

Name _____ Date _____ Class _____

★ **LAB 10-7G** **Use Measurement Tools to Explore Properties of Water**

Use with Lesson 10-7

Materials needed: ruler, tape measure, meter stick, stopwatch, thermometer, scale, plastic container in the shape of a rectangular prism, water
Activity
In this lab you will work with other students to choose measurement tools and use the tools to learn about water.

1. You will need a plastic container that is open on top. Work with a small group of students to guess how much water the container can hold. Your guess should give the weight of the water that the container will hold when it is completely full. Record your guess below. (In the rest of the lab, you will take measurements and find out how good your guess is!)

   Guess: _____

2. Measure the length and width of the container. Record your results below.

   Length of container: _____

   Width of container: _____

   Tool used to measure length and width: _____

3. Calculate the perimeter and area of the top of the container.

   Perimeter of top of container: _____

   Area of top of container: _____

4. Now measure the height of the container.

   Height of container: _____

   Tool used to measure height: _____

5. Calculate the volume of the container.

   Volume of container: _____

6. Slowly fill the container with water. As you do this, keep track of how long it takes to fill the container completely.

   Time it takes to fill container: _____

   Tool used to measure time: _____

Holt Mathematics

TEKS 6.8.B

✪ LAB 10-7G Use Measurement Tools to Explore Properties of Water

7. Find the temperature of the water in the container.

Temperature of water: _____

Tool used to measure temperature: _____

8. Find the weight of the filled container.

Weight of filled container: _____

Tool used to measure weight: _____

9. Pour out the water. Then find the weight of the empty container.

Weight of empty container: _____

Tool used to measure weight: _____

10. Find the weight of the water that was in the container by subtracting the weight of the empty container from the weight of the filled container.

Weight of water: _____

11. How does the actual weight of the water in Step 10 compare to your guess in Step 1?

_____

Try This

Name the tool or tools you would use to measure each of the following.

1. The amount of time it takes to heat a cup of water from 70°F to 80°F

_____

2. The depth of water in a drinking glass

_____

3. The weight of one gallon of water

_____

Holt Mathematics

Name _____ Date _____ Class _____

TEKS 6.8.B

Using Measurement Formulas

LAB 10-7H

Use with Lesson 10-7

Materials needed: ruler or meter stick, stopwatch, thermometer, scale, computer

Activity

The owner's manual for a computer usually includes "specifications" or "specs." The specs give the size, weight, and other information about the computer. Use the following formulas to help you write specs for a computer.

Measurement Formulas		
Perimeter of a Rectangle	$P = 2\ell + 2w$	ℓ = length, w = width
Area of a Rectangle	$A = \ell w$	ℓ = length, w = width
Area of a Triangle	$A = \frac{1}{2}bh$	b = base, h = height
Volume of a Rectangular Prism	$V = \ell wh$	ℓ = length, w = width, h = height
Temperature Conversions	$C = \frac{5}{9}(F - 32)$ $F = \frac{9}{5}C + 32$	C = temperature in degrees Celsius F = temperature in degrees Fahrenheit
Weight Conversions	$Z = 16P$ $P = \frac{Z}{16}$	Z = weight in ounces P = weight in pounds
Time Conversions	$S = 60M$ $M = \frac{S}{60}$	S = time in seconds M = time in minutes

1. Measure the length and width of the computer's screen. Then use formulas to find the perimeter and area of the screen.

   Length of screen: _____

   Width of screen: _____

   Perimeter of screen: _____

   Area of screen: _____

   Formulas used: _____

Holt Mathematics

TEKS 6.8.B

Using Measurement Formulas

LAB 10-7H

2. Measure the length, width, and height of the computer. Then use a formula to find the computer's volume.

Length of computer: _____

Width of computer: _____

Height of computer: _____

Volume of computer: _____

Formula used: _____

3. Use a stopwatch to find out how long it takes to start (or "boot up") the computer. Give the time in seconds. Then use a formula to give the time in minutes.

Time in seconds: _____

Time in minutes: _____

Formula used: _____

4. Carefully place the computer on a scale. Give the weight in pounds. Then use a formula to give the weight in ounces.

Weight in pounds: _____

Weight in ounces: _____

Formula used: _____

5. Computers work best at room temperature. Use a thermometer to find the temperature of the room in degrees Fahrenheit. Then use a formula to give the temperature in degrees Celsius.

Temperature in degrees Fahrenheit: _____

Temperature in degrees Celsius: _____

Formula used: _____

Holt Mathematics

TEKS 6.8.B

⭐ **LAB 10-7H**

Using Measurement Formulas

Try This

Tell which formula you would use to find each of the following.

1. The weight of the keyboard in pounds if you know the weight in ounces

_____

2. The area of a triangular computer stand

_____

3. The temperature of the computer in degrees Fahrenheit if you know the temperature in degrees Celsius

_____

Holt Mathematics

Name _____ Date _____ Class _____

  LAB 12-1 # The Probability of an Event and Its Complement

Use with Lesson 12-1

Materials needed: pencil
Activity
Recall that the probability of an event is the number of ways the event can occur divided by the total number of possible outcomes.

The *complement* of an event is all the ways the event CANNOT happen.

In this lab you will explore the connection between the probability of an event and the probability of its complement.

1. Consider the event that you roll a 1 on a number cube. Complete the table by finding the probability of this event and the probability of its complement.

	Event	**Complement of the Event**
Outcomes	1	2, 3, 4, 5, 6
Probability		

2. Consider the event that you roll an even number on a number cube. Complete the table.

	Event	**Complement of the Event**
Outcomes		
Probability		

3. Consider the event that you roll a number less than 3 on a number cube. Complete the table.

	Event	**Complement of the Event**
Outcomes		
Probability		

Holt Mathematics

TEKS 6.3.B

LAB 12-1 **The Probability of an Event and Its Complement**

For Steps 4-6, use the spinner at right.

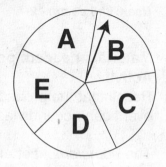

4. Consider the event that you spin an A. Complete the table.

	Event	Complement of the Event
Outcomes		
Probability		

5. Consider the event that you spin a vowel. Complete the table.

	Event	Complement of the Event
Outcomes		
Probability		

6. Consider the event that you spin an A, B, C, or D. Complete the table.

	Event	Complement of the Event
Outcomes		
Probability		

Try This

1. In each table, what do you notice about the probability of the event and the probability of the complement of the event?

_____

2. Describe the relationship between the probability of an event and the probability of its complement.

_____

_____

Holt Mathematics

TEKS 6.9.A

☆ **LAB 12-3** **Constructing Sample Spaces**

Use with Lesson 12-3

Activity

A spinner is divided into thirds and colored red, blue, and green. A second spinner is divided into thirds and colored red, red, and blue. Rosa spins each spinner once.

1. Make a list to find all possible outcomes.

_____

_____

2. Make a tree diagram to find all possible outcomes.

3. How large is the sample space?

_____

Find the number of possible outcomes for each result.

4. red-red _____   **5.** blue-blue _____   **6.** green-green _____

7. Which method (making a list or making a tree diagram) did you prefer for finding all possible outcomes? Why?

_____

_____

_____

Try This

A spinner is divided into fourths and lettered *A-D*. Josh spins the spinner and then tosses a fair number cube.

1. Make a list or a tree diagram to find all possible outcomes.

_____

_____

Holt Mathematics

TEKS 6.9.A

LAB 12-3 **Constructing Sample Spaces**

2. How large is the sample space? _____

Find the number of possible outcomes for each result.

3. A-1 _____   **4.** B-7 _____   **5.** a 6 shows up _____

Holt Mathematics

1-1 Comparing and Ordering Whole Numbers

Values **increase** as you move **right** on a number line.

Values **decrease** as you move **left** on a number line.

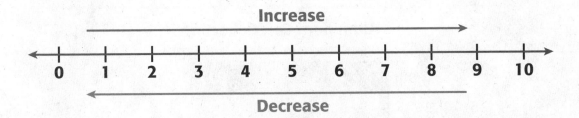

The number line below shows some large whole numbers between 1,000 and 1,100.

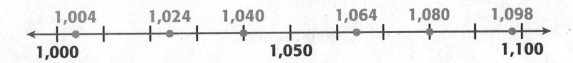

Compare the numbers from the number line above.
Write < or > .

1. 1,024 ☐ 1,080   **2.** 1,024 ☐ 1,004

3. 1,064 ☐ 1,040   **4.** 1,004 ☐ 1,040

5. 1,040 ☐ 1,024   **6.** 1,064 ☐ 1,098

Think and Discuss

7. Explain how to use a number line to compare whole numbers.

8. Describe how to use place value to compare the numbers 1,004 and 1,040.

Holt Mathematics

1-2 Estimating with Whole Numbers

Harvard Middle School is collecting aluminum cans to recycle for a fund-raiser. The number of cans that each grade collected is shown in the bar graph.

Aluminum Can Collection

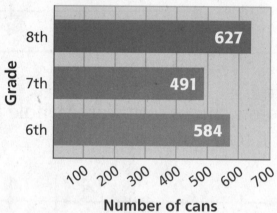

1. Estimate the total number of cans that Harvard Middle School collected.

The principal announced that nearly 2,000 cans were collected. The newspaper reported that over 1,700 cans were collected.

2. Is either of these reports correct?

3. Why are the reports different?

Think and Discuss

4. **Describe** how you reached your estimate.
5. **Identify** some words that indicate whether an amount is an estimate or approximation.

Holt Mathematics

1-3 Exponents

1. Maria won the grand prize on a game show. She will be given $2 the first month, $4 the second month, $8 the third month, and so on as her payment is doubled each month for one year.

   a. Complete the table.

   b. How much will Maria receive in the fifth month?

   c. How much will Maria receive in the eighth month?

   d. Use a calculator to determine how much Maria will receive in the last month of the year.

Month	Amount ($)
1	2
2	4
3	8
4	
5	
6	
7	
8	

Think and Discuss

2. **Describe** the pattern in the table.

3. **Explain** how the values in the table compare with the values 2, 2^2, 2^3, 2^4, and so on.

Holt Mathematics

1-4 Order of Operations

Calculators are programmed to perform operations in a certain order. Each keystroke sequence below results in 17.

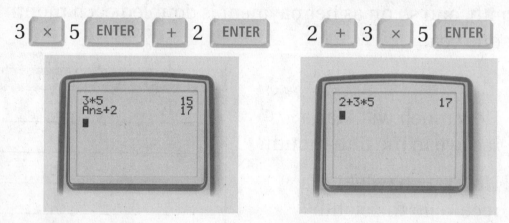

3 [×] 5 [ENTER] [+] 2 [ENTER]   2 [+] 3 [×] 5 [ENTER]

For each keystroke sequence, determine the order of operations the calculator follows.

1. [(] 2 [+] 3 [)] [×] 5 [ENTER]

2. 2 [∧] 3 [−] 1 [×] 4 [ENTER]

3. 2 [∧] [(] 3 [−] 1 [)] [×] 4 [ENTER]

Write the keystroke sequence for each expression.

4. $5 - 2^2$

5. $(2 - 3)^3 + 2$

Think and Discuss

6. **Explain** why there needs to be a rule for the order of operations.

Holt Mathematics

1-5 Mental Math

1. Choose one expression from each pair to evaluate using mental math.

   a. $25 \cdot 24$ $(25 \cdot 4) \cdot 6$

   b. $5 \cdot 22$ $(5 \cdot 20) + (5 \cdot 2)$

   c. $13 + 44 + 27$ $44 + (13 + 27)$

   d. $41 + 32 + 9 + 18$ $(41 + 9) + (32 + 18)$

2. What makes the expressions you chose easier to evaluate?

3. What makes the expressions you did not choose more difficult to evaluate?

Think and Discuss

4. **Discuss** the mental math strategies you used.

5. **Compare** the first expression with the second expression in each pair. How are they alike? How are they different?

Holt Mathematics

1-6 Choose the Method of Computation

Decide whether you would use mental math, pencil and paper, or a calculator to solve each problem. Then solve.

1. Susan makes $9.50 per hour. She worked 7 hours on Monday, 8 hours on Tuesday, 5 hours on Wednesday, and 10 hours on Friday. What is the total amount that Susan earned for the week?

2. Carlos is saving his money to buy a new bike. He earns $45 each week doing yard work, and the bike costs $189. How many weeks will he have to work to have enough money to buy the bike?

3. At a basketball game, 9,980 tickets were sold at $22 each. Find the total amount of money from ticket sales.

4. Rina counted the following numbers of books on each shelf in the storeroom: 24, 47, 26, 53, and 39. Find the total number of books.

5. A group of 12 people wants to rent a room at a pizza restaurant for a party. The room costs $75 to rent. Will $6 from each person be enough to cover the rent?

Think and Discuss

6. **Discuss** when you might choose to use mental math.
7. **Explain** how you decide whether to use pencil and paper or a calculator when you choose not to use mental math.

Holt Mathematics

1-7 Patterns and Sequences

1. Examine the sequence of figures below and look for a pattern.

   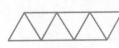

Figure 1   Figure 2   Figure 3   Figure 4   Figure 5   Figure 6

a. Sketch the next two figures in your pattern. Count the number of line segments it takes to draw each.

b. Copy and complete the table for your pattern.

Figure Number	1	2	3	4	5	6	7	8	9	10
Number of Line Segments	3	5	7							

Find the next three numbers in each sequence.

2. 1, 3, 5, 7, ____, ____, ____, …

3. 96, 84, 72, ____, ____, ____, …

4. 1, 3, 6, 10, 15, ____, ____, ____, …

Think and Discuss

5. **Describe** the pattern you noticed in the sequence of triangles.

6. **Explain** how you found the next three numbers in numbers 2–4.

Holt Mathematics

2-1 Variables and Expressions

1. Look at the sequence of connected squares.

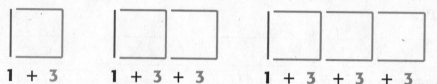

1 + 3   1 + 3 + 3   1 + 3 + 3 + 3

a. Sketch the next two squares.

b. To complete the table for the connected squares, count the number of segments it takes to draw each square.

Number of Connected Squares	1	2	3	4	5	10	20	100
Number of Segments	4	7						

c. How can you find the number of segments if you know the number of squares?

Think and Discuss

2. **Explain** the reasoning you used to find the number of segments in one hundred connected squares.

3. **Explain** the reasoning you could use to find the number of segments in one thousand connected squares.

Holt Mathematics

2-2 Translate Between Words and Math

Drawing pictures and using formulas can help you translate between words and math.

A basketball court is 50 ft wide by 94 ft long. What is its area?

Formula for area

$A = $ length $\times$ width

$A = 94 \times 50$

$A = 4,700$

Area = ?	50 ft

94 ft

The area is 4,700 ft^2.

In some word problems, word order may be confusing. For example, the following problems can be translated in at least two different ways.

Rewrite each problem to make it clearer.

	Word Problem	Possible Translations	Better Word Problem
1.	Write the expression "4 times x plus 6."	$4x + 6$ or $4(x + 6)$	
2.	Translate "the square root of n minus 3."	$\sqrt{n} - 3$ or $\sqrt{n - 3}$	

Think and Discuss

3. Explain what you did to rewrite numbers **1** and **2** to make them easier to translate into math.

Holt Mathematics

EXPLORATION

2-3 Translating Between Tables and Expressions

You can explore geometric patterns to help you write algebraic expressions. Consider this sequence of figures.

Figure 1 **Figure 2** **Figure 3**

The first figure has 5 segments, the second figure has 9 segments, and the third figure has 13 segments.

Figure Number	1	2	3
Number of Segments	5	9	13

1. Draw the next two figures in the pattern.

2. Complete the table.

Figure Number	1	2	3	4	5
Number of Segments	5	9	13		

3. Describe any patterns you notice in the table.

Think and Discuss

4. **Explain** how you can find the number of segments in the 6th figure without drawing it.

5. **Explain** Explain how you can find the number of segments in any figure in the pattern if you know the number of the figure.

Holt Mathematics

2-4 Equations and Their Solutions

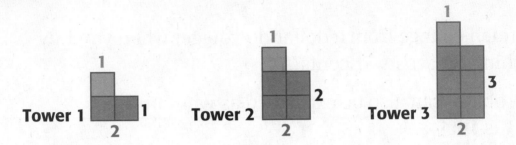

Tower 1   Tower 2   Tower 3

In the sequence of towers, the base of each tower is always 2 squares wide. The heights of the towers vary. If we call the height of each tower h, we can represent this pattern with the following expression:

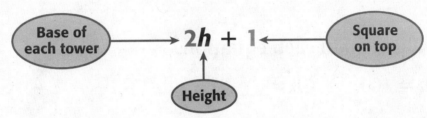

Base of each tower → $2h + 1$ ← Square on top

Height

1. Use the pattern in the sequence of towers to draw a tower with 11 squares. Which tower number is it in the sequence?

2. Use the pattern to solve the equation $2h + 1 = 21$.

3. Look at the sequence of grids and draw a picture of the grid that has 10 shaded squares.

   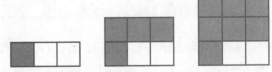

   a. Where in the sequence does this grid occur?

   b. Write an equation for the problem in **3a**.

Think and Discuss

4. **Discuss** what is meant by "a solution of an equation."

Holt Mathematics

2-5 Addition Equations

How much change from a dollar do you get when you buy something that costs 51 cents?

This problem can also be expressed as **what number** plus **51** is **100**?

$$n + 51 = 100$$

$$49 + 51 = 100$$

$$n = 49 \qquad \text{The change is 49¢.}$$

Find the value of n in each equation.

1. $4 + n = 100$ $n =$ _____

2. $n + 45 = 100$ $n =$ _____

3. $19 + n = 100$ $n =$ _____

4. $n + 65 = 100$ $n =$ _____

5. $100 = 41 + n$ $n =$ _____

Think and Discuss

6. **Discuss** your strategies for solving the equations.
7. **Explain** how you can mentally find the solution to $n + 125 = 500$.

Holt Mathematics

2-6 Subtraction Equations

After spending $11, Jane has $3 left in her purse. How much did she have to begin with?

Beginning amount $11 spent   Amount left

n -11 3

$$n - 11 = 3$$
$$14 - 11 = 3$$
$$n = 14$$ She had $14 to begin with.

Find the value of n in each equation.

1. $n - 25 = 75$ $n =$ _____

2. $n - 4 = 19$ $n =$ _____

3. $n - 7 = 35$ $n =$ _____

4. $n - 14 = 21$ $n =$ _____

5. $n - 20 = 83$ $n =$ _____

Think and Discuss

6. **Describe** your strategies for solving the subtraction equations.

7. **Explain** how you can find the solution to $n - 125 = 375$.

Holt Mathematics

2-7 Multiplication Equations

Bill bought 7 tickets to a basketball game for $21. How much did each ticket cost?

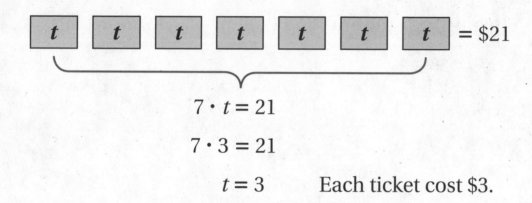

$$7 \cdot t = 21$$
$$7 \cdot 3 = 21$$
$$t = 3 \qquad \text{Each ticket cost \$3.}$$

Find the value of n in each equation.

1. $4 \cdot n = 36$ $n = $_____

2. $100 \cdot n = 500$ $n = $_____

3. $50 = 5 \cdot n$ $n = $_____

4. $24 \cdot n = 48$ $n = $_____

5. $10 \cdot n = 240$ $n = $_____

Think and Discuss

6. Discuss your strategies for solving the equations.

7. Explain how you can find the solution to $4 \cdot n = 200$.

Holt Mathematics

EXPLORATION

2-8 Division Equations

Four friends decided to share the cost of a gift for their dance teacher. After dividing the cost by 4, each friend's share is $25. What was the cost of the gift?

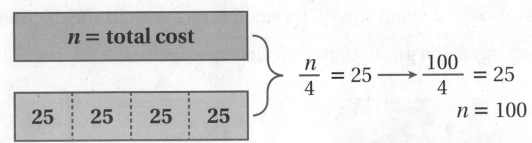

$$\frac{n}{4} = 25 \longrightarrow \frac{100}{4} = 25$$
$$n = 100$$

The gift cost $100.

Find the value of n in each equation.

1. $\frac{n}{2} = 50$

   $n =$ _____

2. $\frac{n}{10} = 2$

   $n =$ _____

3. $20 = \frac{n}{3}$

   $n =$ _____

4. $\frac{n}{7} = 5$

   $n =$ _____

Think and Discuss

5. **Discuss** your strategies for solving the equations.
6. **Explain** how you can find the solution to $\frac{n}{2} = 26$.

Holt Mathematics

3-1 Representing, Comparing, and Ordering Decimals

To model a decimal,

- color a 10-by-10-square grid for each whole in the decimal,
- color one 10-by-1-square strip for each tenth in the decimal, and color a small square for each hundredth in the decimal.

For example, the graph paper models the decimal 1.62.

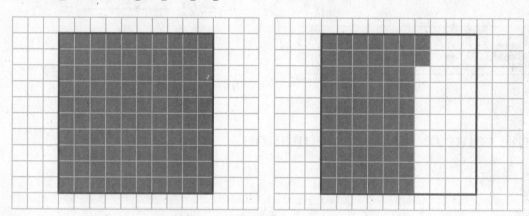

1. Draw a model for each decimal.

   a. 1.25   b. 2.13   c. 1.70   d. 1.7

2. Compare the models for 1c and 1d. What do you notice about these two decimals?

3. Order the decimals in numbers 1a–1d from least to greatest. Explain your reasoning.

Think and Discuss

4. **Explain** why 0.4 = 0.40.

5. **Explain** why 0.5 is greater than 0.10 even though 10 is greater than 5.

Holt Mathematics

3-2 Estimating Decimals

For each problem, estimate a solution. Then compute with a calculator to see how close your estimated solutions are to the actual solutions.

		Estimate	Actual
1.	Seven people want to share the cost of a $33.75 boat rental. How much will each person pay?		
2.	You purchase items that cost the following amounts: $4.95, $1.29, $6.67, $4.19, and $10.39. What is the total cost?		
3.	How much change is there from $200.00 for an item that costs $157.98?		
4.	Ron's gas tank holds 21 gallons and is empty. If gas costs $2.499 per gallon, how much will it cost to fill Ron's tank?		

Think and Discuss

5. **Discuss** the estimation strategies you used.
6. **Describe** a situation in which all you need is an estimated solution and a situation in which you must calculate an exact solution.

Holt Mathematics

3-3 Adding and Subtracting Decimals

You have 19¢. How much more do you need to have $1.00? From 19¢ to 20¢, add 1¢. From 20¢ to $1.00, add 80¢. The answer is 1¢ + 80¢, or 81¢.

1. Draw arrows to connect each pair of amounts that would give you a sum of $1.00.

Amount 1	Amount 2
$0.19	$0.63
$0.25	$0.55
$0.76	$0.93
$0.07	$0.75
$0.65	$0.24
$0.37	$0.35
$0.45	$0.81

2. Compute the change from $10.00 on a purchase of each amount. Example: From $1.25 to $2.00, add $0.75. From $2.00 to $10.00, add $8.00. So the change on a purchase of $1.25 is $0.75 + $8.00 = $8.75.

Amount	Change from $10.00
$1.25	$8.75
$2.76	
$3.07	
$4.65	
$5.37	
$6.45	
$7.59	

Think and Discuss

3. **Name** five different pairs of numbers that each have a sum of $1.00.

4. **Describe** how you can use the strategy of "adding up" to find 200 − 176.25.

Holt Mathematics

3-4 Scientific Notation

You can use exponents to represent powers of 10.

$10^1 = 10$

$10^2 = 10 \times 10 = 100$

$10^3 = 10 \times 10 \times 10 = 1{,}000$

$10^4 = 10 \times 10 \times 10 \times 10 = 10{,}000$

Find each product.

1. 6.25×10^1 **2.** 6.25×10^2 **3.** 6.25×10^3 **4.** 6.25×10^4

In scientific notation, numbers are written as the product of a power of 10 and a number that is greater than 1 and less than 10.

Number	Scientific Notation
6,250	6.25×10^3
62.50	6.25×10^1
625	6.25×10^2

Write each number in scientific notation by filling in the exponent for the power of 10.

5. $42.5 = 4.25 \times 10^{\square}$ **6.** $425 = 4.25 \times 10^{\square}$

7. $4250 = 4.25 \times 10^{\square}$ **8.** $42500 = 4.25 \times 10^{\square}$

Think and Discuss

9. Explain why 5.3×10^3 is greater than 1,000.

10. Explain how you know that $6.25 \times 10^6 = 6{,}250{,}000$.

Holt Mathematics

3-5 Multiplying Decimals

When you multiply decimals, you can use estimation to help you determine the position of the decimal point in the product.

Estimate each product. Then use a calculator to see how reasonable your estimate is.

		Estimate	Actual
1.	4.235 × 16.9		
2.	0.78 × 568		
3.	56.1 × 23		
4.	15.6 × 2.15		

Estimate each product. Use this estimate to decide where to place a decimal point in the answer. Check with your calculator.

5. 70.5 × 4.4 = 3 1 0 2

6. 0.75 × 692 = 5 1 9 0

7. 56 × 3.125 = 1 7 5 0

8. 45.6 × 2.15 = 9 8 0 4

9. 4.17 × 1.2 = 5 0 0 4

10. 125.2 × 7.4 = 9 2 6 4 8

Think and Discuss

11. **Discuss** your strategies for estimating in numbers 1–4.

12. **Explain** how you know where to place the decimal point in a product.

Holt Mathematics

EXPLORATION

3-6 Dividing Decimals by Whole Numbers

1. Four friends go on a vacation together. They decide to share all expenses evenly. Estimate the cost of each item per person, and then compute the actual cost with a calculator.

Item	Total Cost	Estimated Cost per Person	Actual Cost per Person
Cab fare	$50.00		
Pizza	$13.92		
Movie rental	$10.00		
Dinner	$76.20		
Boat ride	$35.96		

Estimate each quotient. Use this estimate to decide where to place a decimal point in the answer. Check with your calculator.

2. $125.2 \div 25 = 5\ 0\ 0\ 8$

3. $40 \div 16 = 2\ 5$

4. $7.5 \div 5 = 1\ 5$

5. $75 \div 12 = 6\ 2\ 5$

Think and Discuss

6. Discuss your strategies for estimating in number **1.**

7. Explain how you know where to place the decimal point in a quotient.

Holt Mathematics

3-7 Dividing by Decimals

A CD store carries the packages of recordable CDs listed in the table. The third column shows the cost of one CD for different packages. This is called the **unit cost.** The fourth column shows how many CDs one dollar can buy. For example, if you buy the 5-pack at $4.95, one dollar buys 1.01 CD (a little more than one CD). This is called the **purchasing power** of $1.00.

Use a calculator to find the cost of 1 CD and the purchasing power of $1.00 for each package.

	Item	Cost	Cost of 1 CD	Purchasing Power of $1.00
1.	Single CD	$1.19		
2.	5-pack	$4.95	$4.95 \div 5 = 0.99$	$5 \div 4.95 = 1.01$
3.	10-pack	$8.95		
4.	20-pack	$16.95		
5.	50-pack	$35.95		

6. Which package gives you the highest unit cost?

7. Which package gives you the lowest unit cost?

8. Which package gives you the greatest purchasing power per dollar? the least purchasing power per dollar?

Think and Discuss

9. Describe the relationship between unit cost and purchasing power by looking at the numbers in the third and fourth columns above.

Holt Mathematics

3-8 Interpret the Quotient

For each problem, estimate a solution. Then compute with a calculator.

		Estimate	Actual
1.	At Juan's school, each lunch special costs $3.65. How many lunches can Juan buy with $20.00?		
2.	Gasoline costs $2.499 per gallon. How many gallons can Sue buy with $25.00?		
3.	On Jorge's map, 0.15 cm represents 1 mi. He measures a road which is 7.8 cm. How many mi long is the actual road?		
4.	Ofelia makes $6.79 per hour at her summer job. If she wants to make $200 per week, how many hours should she work?		

Think and Discuss

5. **Explain** the estimation strategies you used.

6. **Describe** a problem that can be solved by division.

Holt Mathematics

EXPLORATION

3-9 Solving Decimal Equations

For each equation, estimate the solution. Then use
a calculator to solve the equation. Compare the
calculated solution with your estimated solution.

		Estimate	Actual
1.	$1.25 + x = 10$		
2.	$20 - x = 1.95$		
3.	$6x = 15$		
4.	$\dfrac{x}{4.5} = 10$		
5.	$\dfrac{x}{100} = 1.609$		

6. Write a real-world situation for the equation in Exercise **3.**

Think and Discuss

7. **Explain** which equations it was easiest to estimate a solution
for.
8. **Describe** a real-world situation that you could model with a
decimal equation.

Holt Mathematics

4-1 Divisibility

Some calculators have an **INT ÷** key, which returns a quotient and a remainder.

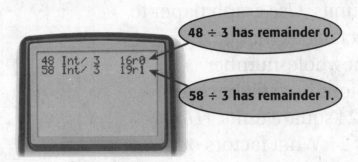

48 ÷ 3 has remainder 0.

58 ÷ 3 has remainder 1.

1. Use mental math or a calculator to determine each quotient and remainder. Then add the digits of the dividend.

	Dividend	Divisor	Quotient	Remainder	Sum of Digits
a.	48	3	16	0	4 + 8 = 12
b.	58	3	19	1	5 + 8 = 13
c.	256	3			
d.	1,011	3			
e.	72	3			
f.	74	3			
g.	129	3			
h.	130	3			

Think and Discuss

2. **Explain** whether 3,129 is divisible by 3.

3. **Describe** the pattern between the remainder and the sum of the digits in the table.

Holt Mathematics

4-2 Factors and Prime Factorization

1. The rectangle measures 4 units by 6 units and has an area of 24 square units. Use graph paper to draw rectangles that have different whole-number dimensions but still have an area of 24 square units. (*Hint:* 4 × 6 = 24. What factors other than 1 × 24 give you 24?)

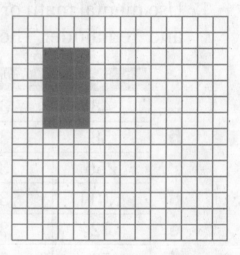

2. The rectangle measures 3 units by 5 units and has an area of 15 square units. Is it possible to draw rectangles that have whole-number dimensions other than 3 × 5 (and 1 × 15) and still have an area of 15 square units?

Think and Discuss

3. **Explain** how you can use rectangles to determine factors of numbers.

4. **Explain** why it is possible to draw more than two different rectangles with an area of 24 square units, but it is not possible to draw more than two different rectangles with an area of 15 square units.

Holt Mathematics

4-3 Greatest Common Factor

The sixth-grade band, which has 60 members, and the seventh-grade band, which has 48 members, are getting ready for a parade. How can they march together in blocks with the same number of columns? The model and table below show one possible formation.

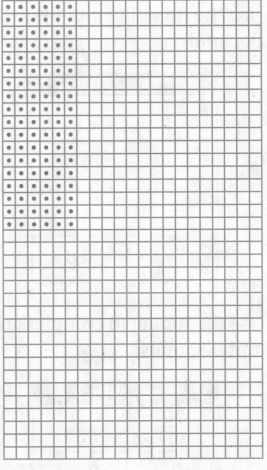

1. Use graph paper to draw a model of two other formations that would work.

2. Complete the table to show the number of rows and columns in the other two formations.

	Formation 1		Formation 2		Formation 3	
	Rows	Columns	Rows	Columns	Rows	Columns
6th Grade	10	6				
7th Grade	8	6				

Think and Discuss

3. **Discuss** which formation the band director should select if she wants the bands to pass through the parade as quickly as possible.

4. **Explain** why both 48 and 60 must be divisible by the number of columns.

Holt Mathematics

4-4 Decimals and Fractions

Use the model to complete the table of equivalent fractions and decimals.

	Fraction	Decimal
1.	$\frac{1}{5}$	
2.		0.6
3.	$\frac{2}{5}$	
4.		0.8

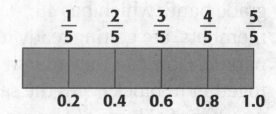

$$\frac{1}{5} \quad \frac{2}{5} \quad \frac{3}{5} \quad \frac{4}{5} \quad \frac{5}{5}$$

0.2   0.4   0.6   0.8   1.0

Use the model to complete the table of equivalent fractions and decimals.

	Fraction	Decimal
5.	$\frac{10}{100}$	
6.		0.45
7.	$\frac{40}{100}$	
8.		0.30

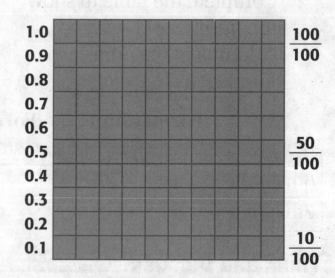

Think and Discuss

9. **Describe** a situation in which decimals are used.

10. **Describe** a situation in which fractions are used.

Holt Mathematics

4-5 Equivalent Fractions

Equivalent fractions are fractions that have the same value.

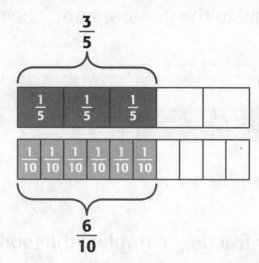

Use the models below to find equivalent fractions.

1. $\frac{1}{2} = \boxed{} = \boxed{} = \boxed{}$

2. $\frac{2}{3} = \boxed{} = \boxed{}$

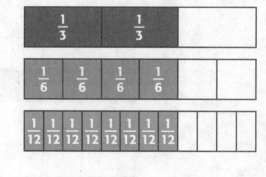

Think and Discuss

3. **Describe** how you could write $\frac{1}{2}$ as an equivalent fraction with a denominator of 24.

4. **Discuss** whether $\frac{6}{12}$ and $\frac{9}{18}$ are equivalent fractions.

Holt Mathematics

4-6 Mixed Numbers and Improper Fractions

An *improper fraction* is a fraction in which the numerator is greater than or equal to the denominator. The model shows that $\frac{7}{4} = \frac{4}{4} + \frac{3}{4} = 1 + \frac{3}{4}$.

The mixed number $1\frac{3}{4} = 1 + \frac{3}{4}$.

For each improper fraction, complete the model and write the improper fraction as a mixed number.

1. $\frac{5}{3}$ 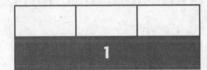 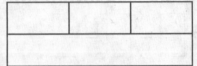 _____

2. $\frac{7}{6}$  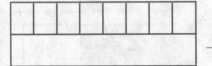 _____

3. $\frac{6}{4}$ 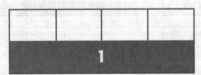 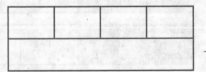 _____

Think and Discuss

4. **Explain** why $\frac{6}{4} = 1\frac{1}{2}$.

5. **Discuss** a situation in which mixed numbers are used.

Holt Mathematics

 **4-7** **Comparing and Ordering Fractions**

Use the model to decide whether the fraction on the left is greater than (>), less than (<), or equal to (=) the fraction on the right.

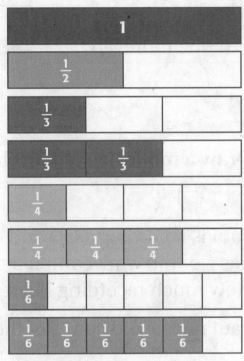

1. $\frac{1}{2} \square \frac{1}{3}$

2. $\frac{1}{2} \square \frac{1}{4}$

3. $\frac{1}{2} \square \frac{2}{3}$

4. $\frac{1}{3} \square \frac{1}{4}$

5. $\frac{3}{4} \square \frac{2}{3}$

6. $\frac{3}{4} \square \frac{5}{6}$

7. $\frac{1}{6} \square \frac{1}{4}$

8. Look at the calculator screen to compare $\frac{1}{2}$ and $\frac{2}{3}$. Which fraction is greater? How do you know?

Think and Discuss

9. **Explain** how you could compare a fraction and a decimal.

10. **Explain** why $\frac{1}{17}$ is greater than $\frac{1}{18}$.

Holt Mathematics

EXPLORATION

4-8 Adding and Subtracting with Like Denominators

Suzanne runs on a $\frac{3}{4}$-mile track. She ran one lap and then decided to run one more lap.

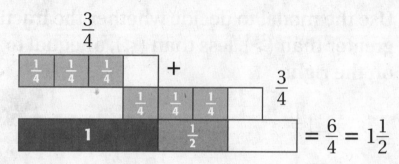

Susan ran $1\frac{1}{2}$ miles.

Draw a model to solve each addition problem.

1. $\frac{3}{5} + \frac{4}{5}$

2. $\frac{3}{10} + \frac{7}{10}$

Paul is recording a CD and has $\frac{1}{6}$ of the work completed. How much recording is left?

Paul needs to record $\frac{5}{6}$ of the CD.

$1 - \frac{1}{6} = \frac{6}{6} - \frac{1}{6} = \frac{5}{6}$

Draw a model to solve each subtraction problem.

3. $\frac{4}{5} - \frac{3}{5}$

4. $1 - \frac{3}{10}$

Think and Discuss

5. **Explain** how to add and subtract fractions with like denominators.

6. **Explain** how to subtract a fraction from 1.

Holt Mathematics

4-9 Estimating Fraction Sums and Differences

Out of 80 students, 49 are in athletics. As the number lines show, approximately half the students are in athletics. In other words, $\frac{49}{80}$ is close to $\frac{1}{2}$.

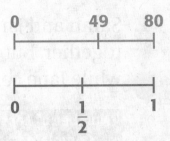

Use a number line to determine whether each fraction is closest to 0, $\frac{1}{2}$, or 1.

1. $\frac{79}{99}$

2. $\frac{22}{213}$

3. $\frac{15}{27}$

4. $\frac{22}{45}$

5. $\frac{300}{475}$

6. $\frac{400}{475}$

Use the estimates you found in numbers 1–6 to estimate each sum or difference.

7. $\frac{79}{99} + \frac{15}{27}$

8. $\frac{22}{213} + \frac{22}{45}$

9. $\frac{300}{475} - \frac{22}{45}$

10. $\frac{15}{27} - \frac{22}{45}$

Think and Discuss _____

11. **Discuss** your strategies for determining whether the fractions were closest to 0, $\frac{1}{2}$, or 1.

12. **Explain** how you know $\frac{237}{475}$ is less than $\frac{1}{2}$.

Holt Mathematics

5-1 Least Common Multiple

1. Sarah and Jane enter a walkathon for charity. They start together, but Sarah completes one lap every **6** minutes while Jane completes one lap every **8** minutes.

Number of Laps Completed	Sarah's Time (min)	Jane's Time (min)
1	$6 \cdot 1 = 6$	$8 \cdot 1 = 8$
2	$6 \cdot 2 = 12$	$8 \cdot 2 = 16$
3	$6 \cdot 3 = 18$	$8 \cdot 3 = 24$
4	$6 \cdot 4 = 24$	$8 \cdot 4 = 32$
5	$6 \cdot 5 = 30$	$8 \cdot 5 = 40$
6	$6 \cdot 6 = 36$	$8 \cdot 6 = 48$
7	$6 \cdot 7 = 42$	$8 \cdot 7 = 56$
8	$6 \cdot 8 = 48$	$8 \cdot 8 = 64$

a. After how many minutes will Sarah and Jane meet at the start again?

b. When will they meet the next time?

Think and Discuss

2. **Discuss** the solution to number **1a** using the term *common multiple.*

3. **Compare** the solution to number **1a** with the solution to number **1b,** and describe these solutions using the terms *common multiple* and *least common multiple.*

Holt Mathematics

5-2 Adding and Subtracting with Unlike Denominators

Fractions are pieces of a whole. When you add or subtract fractions with unlike denominators, you are usually adding or subtracting pieces of different sizes. Look at the model used to solve the problem below.

Phil combines $\frac{1}{4}$ gallon of paint with $\frac{1}{2}$ gallon of paint. How much paint does he have now?

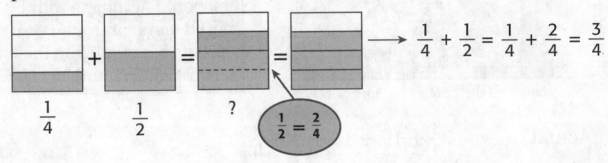

$$\frac{1}{4} + \frac{1}{2} = \frac{1}{4} + \frac{2}{4} = \frac{3}{4}$$

Phil has $\frac{3}{4}$ gallon of paint.

1. Draw a model to show that $1 - \frac{1}{4} = \frac{3}{4}$.

Draw a model to solve each problem. Simplify your answers.

2. $\frac{1}{2} + \frac{1}{3}$

3. $\frac{4}{5} - \frac{1}{2}$

Think and Discuss

4. **Explain** how to add and subtract fractions with unlike denominators.

5. **Draw** a model to show $\frac{1}{2} + \frac{2}{3} = 1\frac{1}{6}$.

Holt Mathematics

5-3 Adding and Subtracting Mixed Numbers

The graph shows typical rainfall levels for a city in the Southwest for the first 5 months of the year. What is the approximate total rainfall from January through May?

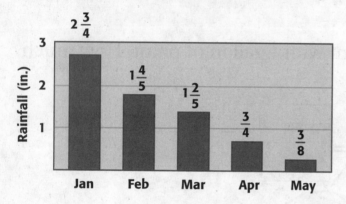

A mixed number contains a whole number and a fraction. To estimate with mixed numbers, round each mixed number to the nearest whole number.

Actual $\qquad 2\frac{3}{4} + 1\frac{4}{5} + 1\frac{2}{5} + \frac{3}{4} + \frac{3}{8}$

Estimated $\qquad 3 + 2 + 1 + 1 + 0 = 7$ in.

The total rainfall is about 7 inches.

Estimate each sum or difference.

1. $13\frac{1}{2} - 9\frac{27}{32}$

2. $1\frac{1}{2} + 9\frac{3}{8} - 2\frac{1}{4}$

3. $17\frac{7}{8} + 19\frac{1}{10}$

4. $4\frac{1}{8} - 1\frac{3}{10} + 3\frac{1}{4}$

Think and Discuss

5. **Discuss** the estimation strategies you used.

6. **Describe** a real-world situation in which mixed numbers are added or subtracted.

Holt Mathematics

5-4 Regrouping to Subtract Mixed Numbers

A baker starts the day with $2\frac{1}{4}$ lemon cakes and sells $1\frac{1}{2}$ lemon cakes during the day. How much cake is left over?

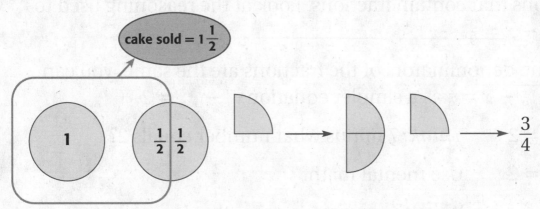

cake sold = $1\frac{1}{2}$

1    $\frac{1}{2}$ $\frac{1}{2}$ $\frac{3}{4}$

initial amount of cake = $2\frac{1}{4}$ cake left = $\frac{3}{4}$

There is $\frac{3}{4}$ of a cake left over.

Use a model to solve each subtraction problem.

1. $2 - 1\frac{1}{4}$

2. $4 - 1\frac{1}{2}$

3. $5\frac{1}{4} - 3\frac{1}{2}$

4. $1\frac{1}{8} - \frac{3}{4}$

Think and Discuss

5. **Discuss** your method for subtracting mixed numbers.

6. **Explain** why the method used to solve the problem about the lemon cakes is called regrouping.

Holt Mathematics

5-5 Solving Fraction Equations: Addition and Subtraction

You can use mental math to solve addition and subtraction equations that contain fractions. Look at the reasoning used to solve $\frac{7}{9} - x = \frac{2}{9}$.

Since the denominators of the fractions are the same, you can rewrite $\frac{7}{9} - x = \frac{2}{9}$ as a simpler equation: $7 - \blacksquare = 2$.

$7 - \blacksquare = 2$	*Think:* 7 minus what number equals 2?
$7 - \boxed{5} = 2$	Use mental math.
$\frac{7}{9} - \frac{5}{9} = \frac{2}{9}$	Write the equation, using the denominator.
$x = \frac{5}{9}$	Write the value of x.

Using the example above as a guide, complete the table below.

	Equation	Simpler Equation	Value of x
1.	$\frac{1}{2} + x = \frac{7}{2}$	$1 + \blacksquare = 7$	
2.	$x - \frac{2}{5} = \frac{1}{5}$	$\blacksquare - 2 = 1$	
3.	$x + \frac{1}{3} = \frac{2}{3}$	$\blacksquare + 1 = 2$	
4.	$x - \frac{1}{3} = \frac{2}{3}$	$\blacksquare - 1 = 2$	

Think and Discuss

5. **Explain** how you could use mental math to solve the equation $\frac{1}{3} + x = \frac{5}{6}$.

Holt Mathematics

5-6 Multiplying Fractions by Whole Numbers

Rosario requires $\frac{3}{4}$ of a 1-pound bag of clay to make one bowl. How many 1-pound bags of clay will she need to make a set of 6 bowls?

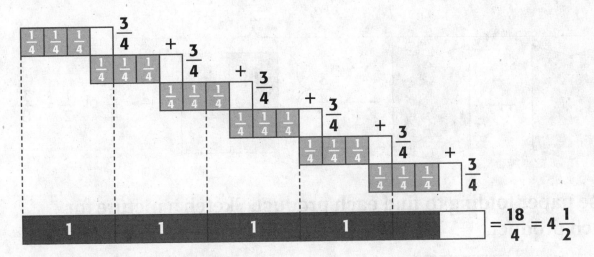

$= \frac{18}{4} = 4\frac{1}{2}$

She will need $4\frac{1}{2}$ 1-pound bags of clay.

Draw a model to find each product.

1. $3 \cdot \frac{3}{4}$

2. $4 \cdot \frac{1}{2}$

3. $5 \cdot \frac{2}{3}$

4. $7 \cdot \frac{1}{4}$

Think and Discuss

5. Explain how you know that $3 \cdot \frac{3}{4}$ is less than 3.

Holt Mathematics

5-7 Multiplying Fractions

You can use paper folding to find the product of two fractions. To find $\frac{3}{4}$ of $\frac{1}{2}$, fold the paper in half vertically to model $\frac{1}{2}$. Then fold it horizontally into four sections to create fourths.

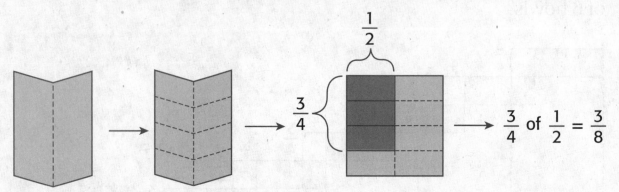

$$\frac{3}{4} \text{ of } \frac{1}{2} = \frac{3}{8}$$

Use paper folding to find each product. Sketch a picture for each product.

1. $\frac{1}{2} \cdot \frac{1}{2}$

2. $\frac{1}{2} \cdot \frac{2}{3}$

3. $\frac{1}{3} \cdot \frac{1}{4}$

4. $\frac{2}{3} \cdot \frac{3}{4}$

Think and Discuss

5. Explain how to multiply two fractions.

Holt Mathematics

5-8 Multiplying Mixed Numbers

You can use paper folding to find products of mixed numbers. To find $\frac{1}{2} \cdot 1\frac{1}{2}$, first fold two sheets of paper in half vertically to represent $1\frac{1}{2}$. To represent $\frac{1}{2}$ of $1\frac{1}{2}$, fold both sheets in half again horizontally.

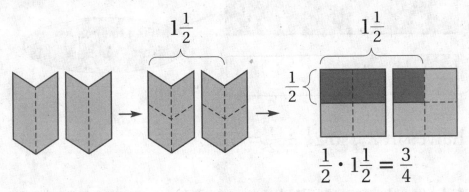

$$\frac{1}{2} \cdot 1\frac{1}{2} = \frac{3}{4}$$

Use paper folding to find each product. Sketch a picture for each product.

1. $\frac{2}{3} \cdot 1\frac{1}{2}$

2. $\frac{1}{3} \cdot 1\frac{3}{4}$

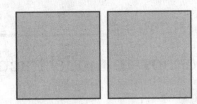

Think and Discuss

3. Explain how to multiply a fraction times a mixed number.

4. Explain why the product of a proper fraction and a mixed number is less than the mixed number.

Holt Mathematics

5-9 Dividing Fractions and Mixed Numbers

The model shows the quotient $2\frac{1}{2} \div \frac{1}{2}$.

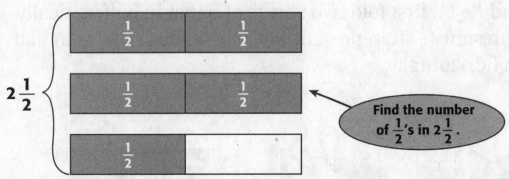

Find the number of $\frac{1}{2}$'s in $2\frac{1}{2}$.

There are 5 halves in $2\frac{1}{2}$, so $2\frac{1}{2} \div \frac{1}{2} = 5$.

Draw a model to solve each division problem.

1. $1\frac{1}{2} \div \frac{3}{4}$

2. $2 \div \frac{2}{3}$

3. $4 \div \frac{2}{3}$

4. $2\frac{1}{4} \div \frac{3}{4}$

Think and Discuss

5. **Describe** how to model fraction division by using fraction bars.

6. **Explain** why $3 \div \frac{3}{4} = 4$.

Holt Mathematics

5-10 Solving Fraction Equations: Multiplication and Division

You can use a number line to solve fraction equations. Look at the reasoning used to solve the equation $\frac{1}{2}n = 50$.

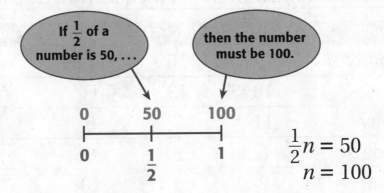

If $\frac{1}{2}$ of a number is 50, ...

then the number must be 100.

$$\frac{1}{2}n = 50$$
$$n = 100$$

Complete the number line to solve each equation.

1. $\frac{1}{2}n = 125$

```
0      125        n
├───────┼─────────┤
0        1        1
         2
```

2. $\frac{1}{3}n = 12$

```
0     12           n
├──────┼─────┬─────┤
0      1     2     1
       3     3
```

3. $\frac{3}{4}n = 60$

```
0            60  n
├───┬───┬────┼───┤
0   1   1   3   1
    4   2   4
```

4. $\frac{2}{3}n = 20$

```
0            20   n
├─────┬──────┼────┤
0     1      2    1
      3      3
```

Think and Discuss

5. **Describe** a real-world situation that could be represented by the equation in number **3**.

6. **Discuss** another way of solving equations that contain fractions and involve multiplication.

Holt Mathematics

6-1 Make a Table

The table shows the number of medals awarded to the top 13 medal-winning countries during the 2002 Winter Olympics.

1. Compute the total number of medals won by each country.

Country	Gold	Silver	Bronze	Total
Germany	12	16	7	
USA	10	13	11	
Norway	11	7	6	
Canada	6	3	8	
Austria	2	4	10	
Russia	6	6	4	
Italy	4	4	4	
France	4	5	2	
Switzerland	3	2	6	
China	2	2	4	
Netherlands	3	5	0	
Finland	4	2	1	
Sweden	0	2	4	

Think and Discuss

2. **Explain** why the table is set up the way it is.

3. **Describe** a different way to organize this data.

Holt Mathematics

6-2 Mean, Median, Mode, and Range

The table shows an ordered list of all the times in the men's 1,500-meter speed-skating competition in the 2002 Winter Olympics. The fastest time was 1:43.95, or 1 minute 43.95 seconds. Notice that all of the times are only seconds apart.

1:43.95	1:44.57	1:45.26	1:45.34	1:45.41	1:45.51	1:45.63	1:45.82
1:45.86	1:45.97	1:45.98	1:46.00	1:46.04	1:46.29	1:46.38	1:46.40
1:46.75	1:46.99	1:47.04	1:47.21	1:47.26	1:47.63	1:47.64	1:47.72
1:47.78	1:47.83	1:48.02	1:48.13	1:48.20	1:48.27	1:48.40	1:48.57
1:48.58	1:48.76	1:49.24	1:49.42	1:49.45	1:49.50	1:49.57	1:50.15
1:50.26	1:50.70	1:51.02	1:51.02	1:51.81	1:52.01	1:52.87	

1. How many seconds behind the winner was the second-place skater? the third-place skater?

2. Find the range. (Subtract the fastest time from the slowest time.)

3. Find the median. (The median is the number in the middle of the data set.)

4. Find the mean. (The mean is the average of all the times.)

Think and Discuss

5. **Discuss** how you found the median.
6. **Explain** how the mean compares with the median.

Holt Mathematics

6-3 Additional Data and Outliers

Casey Fitzrandolph won the men's 500-meter speed-skating competition in the 2002 Olympics with a time of **69.23 seconds.** The table lists the top 32 times in the 500-meter race.

69.23	69.26	69.47	69.49	69.59	69.60	69.60	69.81
69.86	69.89	70.10	70.11	70.28	70.32	70.33	70.44
70.57	70.75	70.84	70.88	70.97	71.27	71.39	71.54
71.96	72.07	72.49	72.58	72.64	72.69	72.93	74.81

1. Find the range, which is the difference between the fastest and the slowest time.

2. Find the median, which is the number in the middle of the data set.

3. Find the mean with a calculator.

4. The table excluded three more times. These times are 108.46, 117.41, and 133.57. Calculate the range, median, and mean including these three additional times.

Think and Discuss

5. **Discuss** why the last three times in number 4 were excluded from the table above.

6. **Describe** how the additional three times affect the range, median, and mean.

Holt Mathematics

6-4 Bar Graphs

The bar graph shows the medal totals of the four countries that won the most medals at the 2002 Winter Olympic Games.

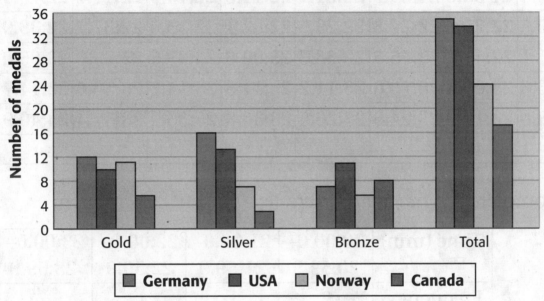

Top Four Medal-Winning Countries at 2002 Winter Olympic Games

1. Name the country that won

   a. the most gold medals.

   b. the most silver medals.

   c. the most bronze medals.

   d. the most medals overall.

2. What is the total number of medals won by each country?

Think and Discuss

3. **Explain** a different way to display the data.

4. **Discuss** information that is not shown by the bar graph.

Holt Mathematics

6-5 Line Plots, Frequency Tables, and Histograms

Below are the top 50 women's times at the 2002 Olympic biathlon.

20:41.4	20:57.0	21:20.4	21:24.1	21:27.9	21:32.1	21:35.7	21:44.2
21:50.3	21:55.6	21:57.0	22:01.7	22:11.9	22:14.9	22:17.7	22:19.7
22:20.6	22:25.8	22:27.3	22:29.9	22:32.1	22:33.5	22:37.7	22:39.9
22:41.1	22:44.7	22:45.5	22:58.3	23:00.0	23:03.5	23:03.8	23:05.0
23:06.6	23:09.4	23:10.0	23:11.2	23:11.3	23:14.2	23:14.6	23:14.7
23:18.0	23:18.9	23:24.6	23:26.5	23:36.8	23:36.9	23:37.4	23:40.9
23:44.1	23:48.7						

1. Complete the *frequency table.*

Time (min)	20:00.0– 20:59.9	21:00.0– 21:59.9	22:00.0– 22:59.9	23:00.0– 23:59.9
Frequency	2			

2. Use the numbers in the frequency table to complete the *histogram.*

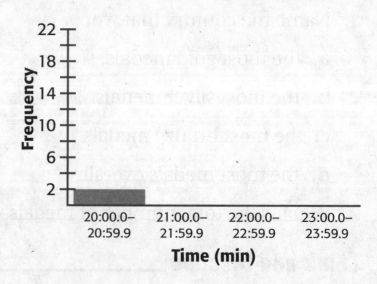

Think and Discuss

3. Explain how you completed the histogram in number 2.

Holt Mathematics

6-6 Ordered Pairs

The table shows the number of faces and vertices of the five regular polyhedrons.

Faces	4	8	20	6	12
Vertices	4	6	12	8	20

This data can be represented on a graph using *ordered pairs.* Each ordered pair is composed of the number of faces and the number of vertices of the polyhedron.

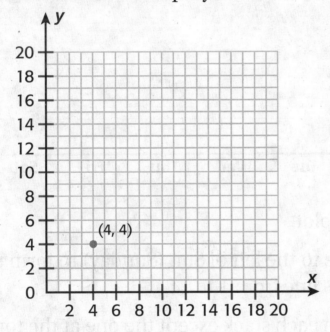

1. Plot and label the remaining ordered pairs. First find the number of faces on the horizontal line. From the number of faces, move up to find the number of vertices on the vertical line.

Think and Discuss

2. **Explain** how to plot ordered pairs.

3. **Discuss** what the point (6, 8) means.

Holt Mathematics

6-7 Line Graphs

Luis works at a record store. His manager asked him to graph the number of CDs returned each day during one week.

Day	Sun	Mon	Tue	Wed	Thu	Fri	Sat
CDs Returned	10	3	2	4	7	11	14

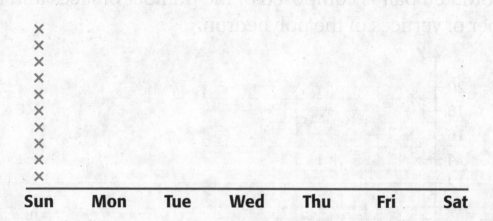

1. Complete the line plot.

2. Draw a vertical line to the left of Sun (Sunday) to form the *y*-axis. Number this line from 1 to 14.

3. Delete all the ×'s in each stack except the one at the top.

4. Connect the ×'s with line segments. You have constructed a *line graph.*

Think and Discuss

5. **Explain** how to construct a line graph.

6. **Discuss** some advantages of displaying data on a line graph rather than in a table.

Holt Mathematics

6-8 Misleading Graphs

The graph shows the total number of medals won by four countries at the 2002 Winter Olympics.

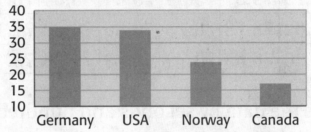

Top Four Medal-Winning Countries at 2002 Winter Olympic Games

1. According to the height of each bar, which country appears to have won approximately half the number of medals won by the United States?

2. Look at the numbers on the left to estimate the number of medals won by each country.

3. Use the estimates in number **2** to determine whether the answer to number **1** is accurate.

Think and Discuss

4. **Discuss** why the graph is misleading.
5. **Explain** how you could modify the graph to represent the data more accurately.

Holt Mathematics

6-9 Stem-and-Leaf Plots

The table shows the times in seconds and hundredths of seconds for the women's 500-meter speed-skating competition in the 2002 Winter Olympics.

74.75	74.94	75.19	75.37	75.39	75.64	75.64	76.17	76.20	76.31
76.37	76.42	76.62	76.73	76.73	76.86	76.92	77.10	77.37	77.60
77.60	77.71	78.26	78.63	78.79	78.89	79.28	79.45	79.45	

You can organize the data by seconds and hundredths of seconds. Notice how the times are grouped using different colors.

1. Complete the *stem-and-leaf plot.* The stems represent seconds and the leaves represent hundredths of seconds. Notice how the colors correspond to the colors used in the table.

Stems	Leaves
74	75 94
75	
76	
77	
78	
79	

Key: 74 | 75 means 74.75

Think and Discuss

2. **Explain** what it means in this case for a stem to have the most number of leaves.

3. **Explain** what it means in this case for a stem to have the least number of leaves.

Holt Mathematics

6-10 Choosing an Appropriate Display

The graphs shown below are not labeled. Match each of the descriptions with the graph that most likely represents the data.

1. The population of a town over the course of several years

2. The areas of the five Great Lakes

3. The test scores of the students in a Spanish class

Graph A

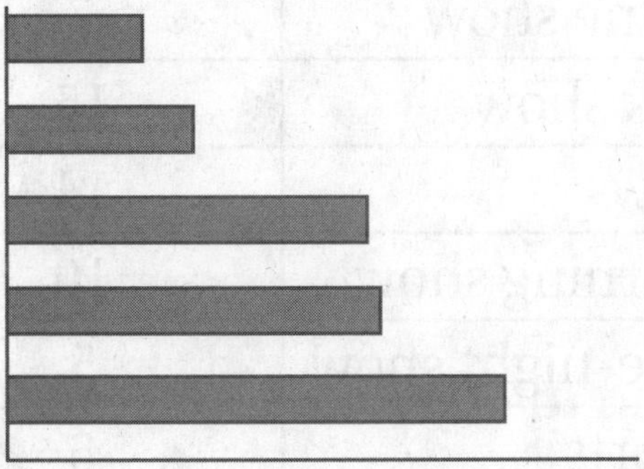

Graph B

Stems	Leaves
6	3 4 7
7	5 5 5 8
8	2 3 5 7 9 9
9	1 4 8

Key: 6 | 3 means 63

Graph C

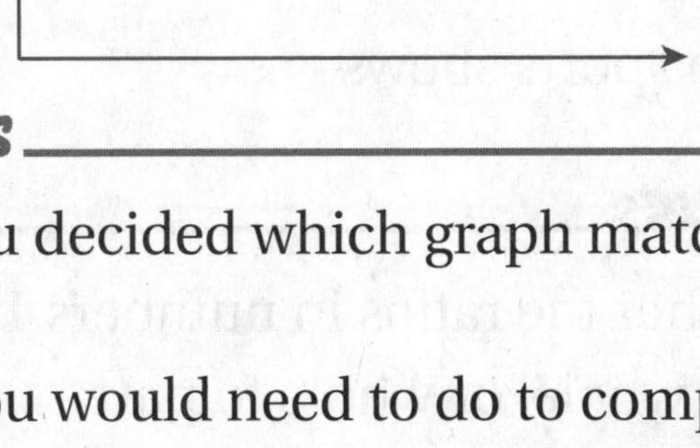

Think and Discuss

4. **Explain** how you decided which graph matches each description.

5. **Explain** what you would need to do to complete Graph A.

Holt Mathematics

7-1 Ratios and Rates

A TV network offers the numbers of shows each week shown in the table.

You can compare the numbers of TV shows by using ratios. A **ratio** is a comparison of two quantities that uses division. For example, the ratio of science fiction shows to drama shows is $\frac{3}{14}$, which can also be written 3:14 or 3 to 14.

Type of TV Show	Number of Shows
Comedy	14
Drama	14
Science fiction	3
Game show	7
Talk show	15
News	14
Morning show	10
Late-night show	5
Sports	6

Find each ratio.

1. comedy shows to game shows

2. game shows to news shows

3. morning shows to late-night shows

4. talk shows to sports shows

Think and Discuss

5. **Discuss** whether the ratios in numbers 1–4 compare part to part, part to whole, or whole to part.

6. **Discuss** whether order is important when calculating ratios. (*Hint:* Is $\frac{news}{sports}$ equivalent to $\frac{sports}{news}$?)

Holt Mathematics

7-2 Using Tables to Explore Equivalent Ratios and Rates

McMillans Restaurant is celebrating its 50th anniversary by offering 3 hamburgers for $2.

1. Sue makes this table for quick reference at the drive-up window. Complete the table.

Number of Hamburgers	3	6	9		15		21
Total Cost ($)	2	4					

2. The weekend goal is to sell 1,200 hamburgers. How much money will the restaurant receive if it reaches its goal of 1,200 hamburgers? Complete the table to help you answer this question.

Number of Hamburgers Sold	300	600	900	1,200
Amount Received ($)	200			

Think and Discuss

3. Describe any patterns you notice in the table for Exercise 1.

4. Explain how you could find the amount of money the restaurant would receive if it sold 1,500 hamburgers.

Holt Mathematics

7-3 Proportions

An automobile assembly line finishes 3 cars every 2 hours.

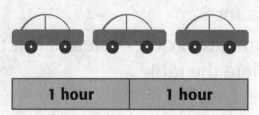

1 hour	1 hour

1. Use the diagram to determine how many cars are finished each hour.

2. Use the diagram to determine approximately how long it takes to finish 1 car.

3. If it takes 2 hours to finish 3 cars, how many hours does it take to finish

   a. 6 cars?   b. 9 cars?   c. 12 cars?

   d. 4 cars?   e. 8 cars?   f. 16 cars?

Think and Discuss

4. **Discuss** how you used the diagram to solve numbers 1 and 2.

5. **Explain** how you solved numbers 3a–3f.

Holt Mathematics

7-4 Similar Figures

Similar rectangles have the same shape but may be different sizes.

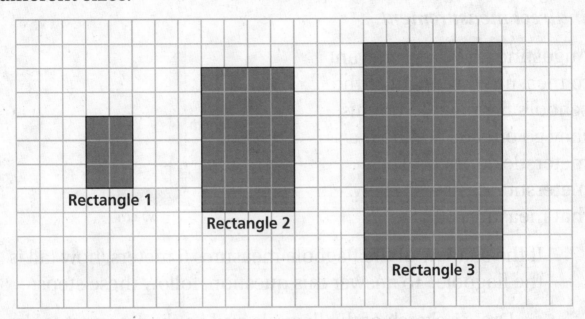

Rectangle 1

Rectangle 2

Rectangle 3

Determine the length and width of each rectangle and find each ratio.

Ratio 1	Ratio 2
1. $\dfrac{\text{length of rectangle 1}}{\text{length of rectangle 2}} =$	$\dfrac{\text{width of rectangle 1}}{\text{width of rectangle 2}} =$
2. $\dfrac{\text{length of rectangle 2}}{\text{length of rectangle 3}} =$	$\dfrac{\text{width of rectangle 2}}{\text{width of rectangle 3}} =$
3. $\dfrac{\text{length of rectangle 1}}{\text{length of rectangle 3}} =$	$\dfrac{\text{width of rectangle 1}}{\text{width of rectangle 3}} =$

Think and Discuss

4. Describe the pattern between ratio 1 and ratio 2.

5. Explain why the three rectangles are similar.

Holt Mathematics

7-5 Indirect Measurement

The heights of very tall structures can be measured indirectly using similar figures and proportions. This method is called *indirect measurement.*

Augustine and Carmen want to measure the height of the school's flagpole. To do this, they go outside and hold a meterstick upright. The meterstick casts a shadow that measures 50 cm.

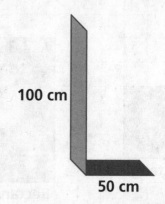

100 cm

50 cm

1. If the shadow of the flagpole measures 6 meters, how tall is the flagpole? To answer this question, follow these steps:

   a. Draw a sketch of the flagpole and its shadow next to the sketch of the meterstick and its shadow.

   b. Label the height of the flagpole *x*.

   c. Write the proportion $\frac{\text{height of flagpole}}{\text{shadow of flagpole}} = \frac{\text{height of meterstick}}{\text{shadow of meterstick}}$, and substitute the values for the given measurements.

   d. Solve the proportion for *x*.

Think and Discuss

2. **Explain** how you solved the proportion for *x* in number **1d.**

3. **Explain** whether you could have solved the problem by writing the proportion as $\frac{\text{shadow of flagpole}}{\text{height of flagpole}} = \frac{\text{shadow of meterstick}}{\text{height of meterstick}}$.

Holt Mathematics

7-6 Scale Drawings and Maps

A *scale* is a ratio between two sets of measurements. For example, the scale 2 in:1 mi means that 2 inches on a scale drawing represents 1 mile.

1. Each letter of the Hollywood sign measures 50 ft tall and 30 ft wide. Use the rectangles below to sketch a scale drawing of the Hollywood sign. The side lengths of each square inside each rectangle represent 10 ft.

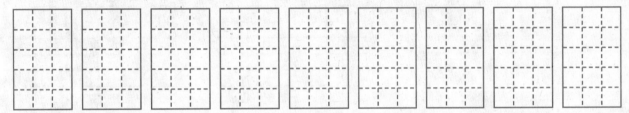

2. If the total width of the Hollywood sign is approximately 450 ft, what is the approximate distance between each pair of neighboring letters?

Think and Discuss

3. **Explain** how you found the answer in number 2.
4. **Discuss** other examples of scale drawings.

Holt Mathematics

7-7 Percents

Percent means "per one hundred." The decimal grid shows 50%, or 50 out of 100.

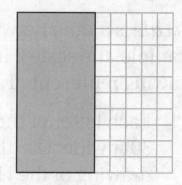

Use the decimal grid to show a model of each percent. Then write the percent as a fraction in simplest form.

1. 25% = _____   **2.** 75% = _____   **3.** 80% = _____

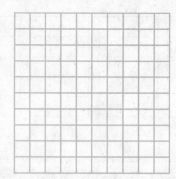   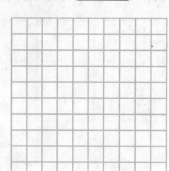   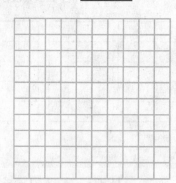

Determine the percent modeled by each decimal grid, and then write it as a fraction in simplest form.

4.   **5.**   **6.**

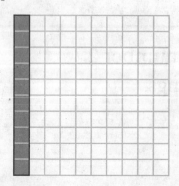   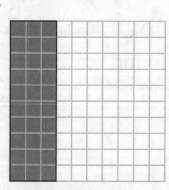   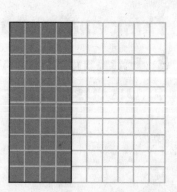

Think and Discuss

7. Explain how to write a percent as a fraction.

Holt Mathematics

7-8 Percents, Decimals, and Fractions

To report what percent of their fund-raising goal has been reached, a charity uses the number-line model below.

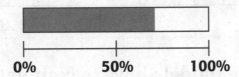

1. Has the charity reached about 50%, about 75%, or about 100% of its goal?

Complete each number-line model by writing percents above the line and the corresponding fractions below the line.

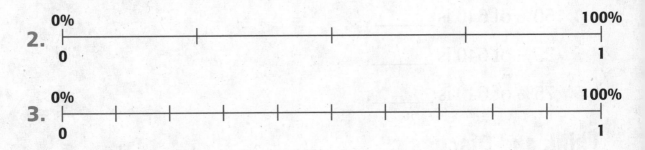

2.

3.

Think and Discuss

4. Explain how you matched percents with fractions in numbers **2** and **3**.

5. Explain how you could label the number lines with decimals.

Holt Mathematics

7-9 Percent Problems

You can use a number line to find the percent of a number.

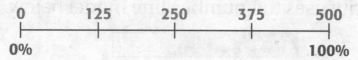

Use the number-line model above to complete each problem.

1. $\dfrac{125}{500} =$ _____% **2.** $\dfrac{250}{500} =$ _____% **3.** $\dfrac{375}{500} =$ _____%

Label the number-line model above to find each percent of 640.

4. 50% of 640 is _____.

5. 25% of 640 is _____.

6. 75% of 640 is _____.

Think and Discuss

7. Discuss what it means to find 100% of a number.
(*Hint:* What is 100% of 640?)

8. Explain how you can use number-line models to solve percent problems.

Holt Mathematics

7-10 Using Percents

Stores that go out of business often offer big discounts on purchases. In such situations, 50% off sales are common.

Estimate the discount for each item at 50% off. Then calculate the actual discount.

	Item	Price	Estimated Discount	Actual Discount
1.	Shirt	$39.95		
2.	DVD player	$288.99		
3.	Speakers	$239.95		
4.	TV	$1,035.29		
5.	MP3 player	$247.99		

Think and Discuss

6. **Discuss** the estimation strategies you used.

7. **Explain** whether a one-time 50% discount is equivalent to two consecutive 25% discounts. (*Hint:* Use $100.00 as the base amount.)

Holt Mathematics

8-1 Building Blocks of Geometry

Geometry can be used to describe the physical world around us.
Check the box of the geometry term that each real-world item
represents.

		Point	Line Segment	Plane
1.	A freckle			
2.	A strand of hair			
3.	A poster			
4.	A pixel on your calculator screen			
5.	A period at the end of a sentence			
6.	A guitar string			
7.	The minute hand of a clock			
8.	A computer screen			

Think and Discuss

9. **Describe** the characteristics of the items that you classified
   as *points* in the table above.
10. **Describe** the characteristics of the items that you classified
    as *line segments* in the table above.

Holt Mathematics

8-2 Measuring and Classifying Angles

An angle is formed by two rays that have a common endpoint. Right angles measure 90° and are shaped like a letter **L**. You can estimate an angle measure by comparing the angle with a right angle.

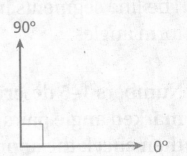

Estimate the measure of each angle. Then measure the angle with a protractor.

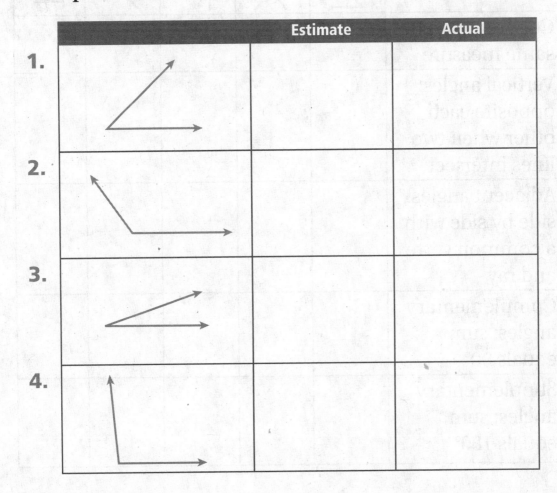

		Estimate	Actual
1.			
2.			
3.			
4.			

Think and Discuss

5. Discuss how you estimated the angle measures.

Holt Mathematics

8-3 Angle Relationships

The line segments in some letters, symbols, and numbers form angles.

Numbers 1–5 describe types of angle pairs. Determine which marked angle pairs in the figures apply to each description, and then check the appropriate boxes.

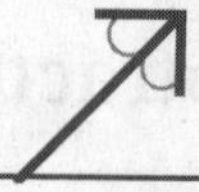

		Z	X	F	↗	4
1.	Congruent angles: same measure					
2.	Vertical angles: opposite each other when two lines intersect					
3.	Adjacent angles: side by side with a common vertex and ray					
4.	Complementary angles: sum equals 90°					
5.	Supplementary angles: sum equals 180°					

Think and Discuss

6. Discuss other examples of vertical angles in the real world.

Holt Mathematics

8-4 Classifying Lines

The table shows pairs of intersecting lines, pairs of parallel lines, and pairs of perpendicular lines.

Intersecting Lines	Parallel Lines	Perpendicular Lines

1. Draw your own examples of intersecting lines, parallel lines, and perpendicular lines.

Think and Discuss

2. **Describe** how you can tell if two lines are parallel.

3. **Discuss** what makes perpendicular lines different from other lines that intersect.

Holt Mathematics

8-5 Triangles

You can classify triangles by the measures of their angles.

Measure the angles of each triangle. Then check the box that gives the correct classification of the triangle.

		Acute Triangle: has only acute (less than 90°) angles	Obtuse Triangle: has one obtuse (greater than 90°) angle	Right Triangle: has one right (90°) angle
1.				
2.				
3.				

Think and Discuss

4. **Find** the sum of the angle measures in each triangle.
5. **Make** a generalization about the sum of the angle measures in a triangle.

Holt Mathematics

8-6 Quadrilaterals

Find a real-world example for each quadrilateral.

	Quadrilateral	Example
1.	Parallelogram	
2.	Rhombus	
3.	Rectangle	
4.	Trapezoid	

Think and Discuss

5. **Discuss** how all the quadrilaterals are similar.

6. **Explain** what is special about the rectangle.

Holt Mathematics

8-7 Polygons

In a *regular polygon,* all sides are congruent and all angles are congruent.

Name each polygon and determine whether it is regular. Use number 1 as an example.

	Polygon	Name	Regular?
1.		Triangle	no
2.			
3.			
4.			
5.			
6.			

Think and Discuss

7. **Explain** how you classified each polygon in numbers 2–6.

Holt Mathematics

8-8 Geometric Patterns

Look for a pattern, and draw the next three figures in the sequence.

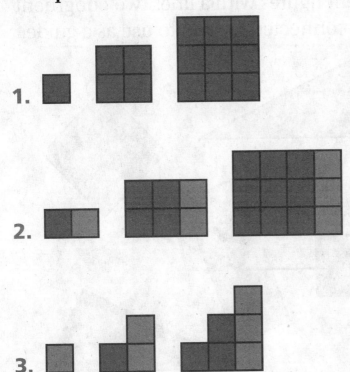

1.

2.

3.

Think and Discuss

4. **Explain** how the sequence in number 2 is built on the sequence in number 1.
5. **Describe** in words the sequence in number 3.

Holt Mathematics

8-9 Congruence

Congruent figures are exactly the same shape and size.

1. Connect two congruent figures with a line. Two congruent rectangles have been connected for you to use as a guide.

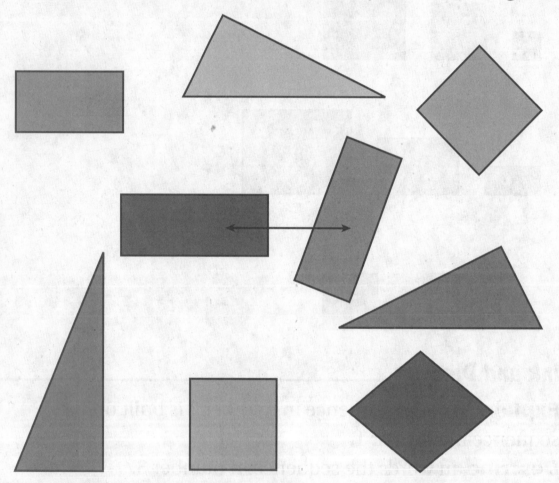

2. Measure the sides and angles of each pair of connected figures to be sure that they are congruent.

Think and Discuss

3. **Give examples** of congruent figures that occur in the real world.

Holt Mathematics

8-10 Transformations

The green triangle is a *reflection* of the blue triangle across the solid vertical line.

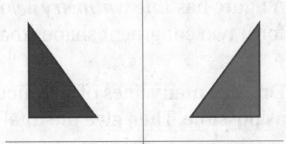

1. Reflect the green triangle across the horizontal line, and then reflect the resulting triangle across the vertical line.

The orange triangle is a *translation* of the yellow triangle down and to the right.

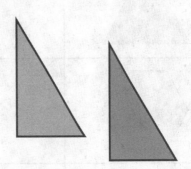

2. Translate the orange triangle up and to the right.

Think and Discuss

3. **Define** *reflection* in your own words.
4. **Define** *translation* in your own words.

Holt Mathematics

8-11 Line Symmetry

A figure has *line symmetry* if you can draw a line through it to form two congruent shapes that are reflections of each other.

Draw as many lines of symmetry through each figure as possible. Then give the total number of lines drawn in each.

	How Many Lines of Symmetry?
1.	
2.	
3.	
4.	

Think and Discuss

5. **Discuss** the characteristics of the figures that have more than one line of symmetry.

6. **Discuss** the characteristics of the figures that have only one line of symmetry.

Holt Mathematics

9-1 Understanding Customary Units of Measure

Different units of measure are used for measuring length, weight, and capacity. Capacity is the amount that a container can hold.

1. Decide whether each of the units of measure below is used to measure length, weight, or capacity. Write the name of the unit in the appropriate column of the table.

Pint   Pound   Inch   Ton

Foot   Quart   Mile   Gallon

Yard   Ounce   Cup   Fluid Ounce

Length	Weight	Capacity

Think and Discuss

2. **Describe** an object that could be measured using yards.

3. **Explain** how you would decide whether to measure an object using inches or feet.

Holt Mathematics

9-2 Understanding Metric Units of Measure

In the metric system, the basic unit of length is the meter. One meter (1 m) is about the width of a classroom doorway. Other metric units of length are based on the meter.

Unit	Fraction of a Meter	Decimal Part of a Meter
Millimeter	$\frac{1}{1,000}$ of a meter	0.001 m
Centimeter	$\frac{1}{100}$ of a meter	0.01 m
Decimeter	$\frac{1}{10}$ of a meter	0.1 m

1. Which of the units listed is the smallest? How do you know?

2. Which of the units listed is the largest? How do you know?

3. In the metric system, the basic unit of capacity is the liter (L). The prefixes *milli-*, *centi-*, and *deci-* are used in the same way to create other units of capacity. Complete the table.

Unit	Fraction of a Liter	Decimal Part of a Liter
Milliliter	__?__ of a liter	__?__
Centiliter	__?__ of a liter	__?__
Deciliter	__?__ of a liter	__?__

Think and Discuss

4. **Explain** which is longer, a stick that is 5 centimeters long or a stick that is 5 decimeters long.

5. **Explain** which holds more, a container with a capacity of 7 milliliters or a container with a capacity of 7 centiliters.

Holt Mathematics

9-3 Converting Customary Units

One foot is equal to 12 inches. You can use this fact to explore the relationship between feet and inches.

1. Complete the table.

Feet	1	2	3	4	5	6	7	8
Inches	12	24						

2. Complete the table.

Feet								
Inches	108	120	132	144	156	168	180	192

3. If you are given a length in feet, how can you convert the length to inches?

4. If you are given a length in inches, how can you convert the length to feet?

Think and Discuss

5. Describe what you would do to convert 324 inches to feet.

6. Explain how you could use the above tables to convert 3.5 feet to inches.

Holt Mathematics

9-4 Converting Metric Units

Because the metric table is based on powers of 10, it is important to be able to multiply and divide quickly by powers of 10.

1. Complete the table by multiplying by 10, by 100, and by 1,000. You may use a calculator or any other method you wish. Look for patterns as you work.

Number	× 10	× 100	× 1,000
42			
3.8			
0.97			
0.065			

2. Complete the table by dividing by 10, by 100, and by 1,000. You may use a calculator or any other method you wish. Look for patterns as you work.

Number	÷ 10	÷ 100	÷ 1,000
512			
63.9			
4.05			
0.772			

Think and Discuss

3. **Describe** shortcuts for multiplying by 10, by 100, and by 1,000.

4. **Explain** shortcuts for dividing by 10, by 100, and by 1,000.

Holt Mathematics

9-5 Time and Temperature

One hour is divided into 60 minutes. You can use proportional thinking to explore the relationship between hours and minutes.

1. How many minutes are there in 2 hours? in 3 hours?

2. How many minutes are there in $\frac{1}{2}$ hour? in $\frac{1}{4}$ hour?

3. Complete the table.

Hours	$\frac{1}{10}$	$\frac{1}{5}$	4	6		
Minutes					480	600

4. What fraction of an hour is 2 minutes?

5. In general, how can you convert any number of minutes to hours?

6. In general, how can you convert any number of hours to minutes?

Think and Discuss

7. **Explain** how to write 23 minutes as a fraction of an hour.

8. **Discuss** how you could write 72 minutes in terms of hours and minutes.

Holt Mathematics

9-6 Finding Angle Measures in Polygons

The figures show the measures of several angles.

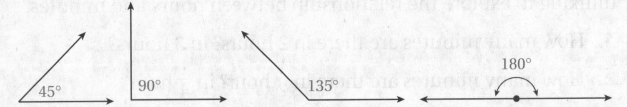

45° 90° 135° 180°

You can use these angles as benchmarks to help you estimate the measures of other angles.

Estimate the measure of each angle.

1.

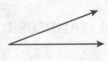

2.

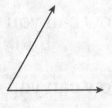

3.

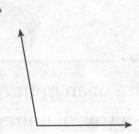

4.

Think and Discuss

5. Explain how you could use the corner of an index card to help you estimate angle measures.

6. Describe how you could use the corner of an index card to draw a 135° angle. (*Hint:* Consider folding the corner of the card.)

Holt Mathematics

9-7 Perimeter

Felicia is working with her dad to design a deck for their yard. They sketch the floor space on grid paper with the side length of each square representing 2 feet.

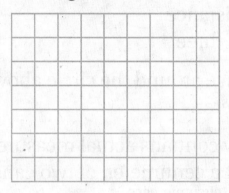

1. Label the dimensions in feet of the deck.

2. The final task in building the deck is to nail a trim piece all the way around the deck. Find the distance around the deck.

3. The distance around the deck is called the *perimeter.* What are three other real-world situations in which you might want to find the perimeter?

Think and Discuss

4. **Discuss** your method for finding the perimeter of the deck.

5. **Explain** how you could write a formula for perimeter of a rectangle using ℓ for length and w for width.

Holt Mathematics

9-8 Circles and Circumference

People in ancient civilizations
learned to estimate the distance
around a circle (***circumference***)
by multiplying the distance
across (***diameter***) by three.

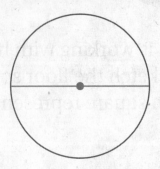

1. Does the distance around the circle above look like three
    times the diameter?

2. The table below contains actual measurements, to the
    nearest tenth of a centimeter, of two cans. Use a calculator
    to find the ratio $\frac{\text{circumference}}{\text{diameter}}$.

Object	Diameter (cm)	Circumference (cm)	Circumference Diameter
Juice can	5.2	16.3	
Coffee can	15.7	49.3	

3. Are the $\frac{\text{circumference}}{\text{diameter}}$ ratios you found in number **2** reasonably
    close to 3?

Think and Discuss

4. **Discuss** whether the ratio $\frac{\text{circumference}}{\text{diameter}}$ changes according
    to the size of the circle.

Holt Mathematics

10-1 Estimating and Finding Area

Mr. and Mrs. Domínguez want to have the bottom of their pool refinished. A sketch of the pool is shown below with the side length of each square representing 1 yard. Before they begin the refinishing project, they have to estimate the area in square yards of the bottom of the pool.

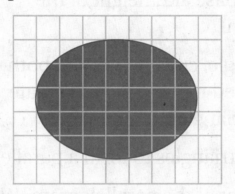

1. Estimate the area, in square yards, of the bottom of the pool.

2. Compare your estimate with the estimates of others in your class, and then average your estimates.

3. What are three other real-world situations in which you might want to estimate area?

Think and Discuss

4. **Discuss** the strategies you used for estimating the area of the bottom of the pool.

5. **Explain** how you could use squares to help you estimate the areas of irregular shapes.

Holt Mathematics

10-2 Area of Triangles and Trapezoids

You can use what you know about the area of a parallelogram to develop a formula for the area of a triangle.

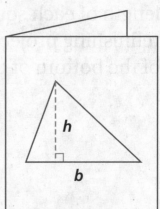

1. Fold a sheet of paper in half. Use a ruler to draw a triangle on one side of the folded sheet. Label the base and height of the triangle as shown.

2. Cut out the triangle, cutting through both layers of the folded paper. This will create two congruent triangles.

3. Arrange the two triangles to form a parallelogram.

4. What is the height of the parallelogram? What is its base?

5. What is the area of the parallelogram?

6. How is the area of the triangle related to the area of the parallelogram?

Think and Discuss

7. **Show** how to write a formula for the area of a triangle with base b and height h.

8. **Explain** how to use your formula to find the area of this triangle.

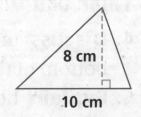

Holt Mathematics

10-3 Area of Composite Figures

Phil and Louise are planning to sod their backyard. Sod is sold in square yards. They sketched their yard on a piece of graph paper, where the side length of each square represents 1 yard.

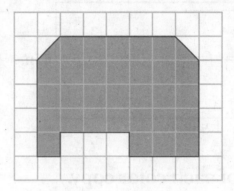

1. **a.** Divide the figure into several simpler figures.

   b. Name each figure from part **a.**

   c. Find the area in square yards of each figure.

   d. Add the areas in part **c.**

Think and Discuss

2. **Discuss** the strategies you used for finding area.

3. **Explain** how you could use squares to help you find the areas of irregular shapes.

Holt Mathematics

10-4 Comparing Perimeter and Area

Suzanne is enlarging a color copy of a 3 in. by 5 in. photograph to 6 in. by 10 in. A model is shown below.

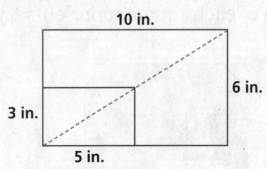

Find the perimeter of each color copy.

	Color Copy	Perimeter = 2 · Length + 2 · Width
1.	3 in. by 5 in.	
2.	6 in. by 10 in.	

3. How do the perimeters compare?

Find the area of each color copy.

	Color Copy	Area = Length · Width
4.	3 in. by 5 in.	
5.	6 in. by 10 in.	

6. How do the areas compare?

Think and Discuss

7. Explain how you compared the perimeters.

8. Explain how you compared the areas.

Holt Mathematics

10-5 Area of Circles

You can use estimation to help you investigate the area of circles.

1. Estimate the area of the circle by counting whole and partial squares.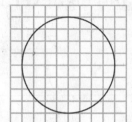

2. What is the radius *r* of the circle?

3. Calculate πr^2 using 3.14 for *pi*.

4. How does the value of πr^2 compare to your estimate of the circle's area?

5. Repeat the process for this circle. First estimate the area by counting whole and partial squares.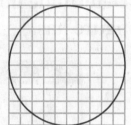

6. What is the radius of the circle?

7. Calculate πr^2 using 3.14 for *pi*.

8. How does the value of πr^2 compare to your estimate of the circle's area?

Think and Discuss

9. **Describe** any shortcuts you found for counting the squares.

10. **Explain** how you can use what you discovered to write a formula for the area of a circle with radius *r*.

Holt Mathematics

10-6 Three-Dimensional Figures

A cube is a solid figure with six faces, twelve edges, and eight vertices.

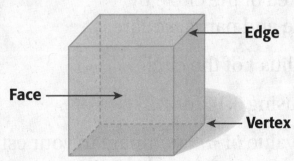

Determine how many faces, edges, and vertices each solid figure has.

		Faces	Edges	Vertices
1.				
2.				
3.				

Think and Discuss _____

4. Explain how many edges it takes to form a vertex.

5. Explain how many faces it takes to form an edge.

Holt Mathematics

10-7 Volume of Prisms

Volume is the number of cubic units that fill a space. Notice how the volume of a rectangular prism increases as the height increases.

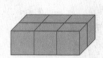   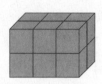   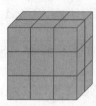

$V = 6$ cubic units $V = 12$ cubic units $V = 18$ cubic units

Find the volume of each rectangular prism

1.

2.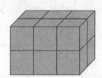

3.

4.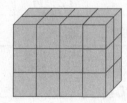

Think and Discuss

5. **Explain** how you found the volume of each rectangular prism.

6. **Discuss** why the formulas $V = $ base $\cdot$ height and $V = $ length $\cdot$ width $\cdot$ height are equivalent.

Holt Mathematics

10-8 Volume of Cylinders

The area of the base of a soup can is 4.9 in^2, and the height is 4 in. To find the volume of this can, multiply the area of the base times the height.

volume = area of base · height ($V = Bh$)

$V = 4.9 \cdot 4 = 19.6$ in^3

The soup can has a volume of 19.6 in^3.

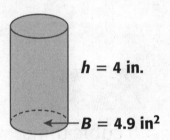

$h = 4$ in.

$B = 4.9$ in^2

Find the volume of each cylinder.

	Area of Base	Height	Volume = Area of Base · Height
1.	12.6 in^2	8 in.	
2.	28.3 cm^2	10 cm	
3.	3.14 ft^2	2 ft	
4.	113.1 in^2	12 in.	
5.	176.7 cm^2	25 cm	

Think and Discuss

6. **Explain** how to find the volume of a cylinder.
7. **Discuss** why the formulas $V = B \cdot h$ and $V = \pi \cdot r^2 \cdot h$ are equivalent ($r =$ radius).

Holt Mathematics

10-9 Surface Area

You can use grid paper to make nets that cover boxes, or rectangular solids. The area of the net is the *surface area of the solid.*

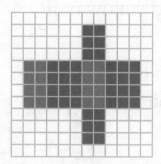

1. Find the combined area of the blue rectangles (the sides of the box).

2. Find the combined area of the green rectangles (the top and bottom of the box).

3. Add the areas you found in numbers **1** and **2**. This is the surface area of the box.

4. On the grid below, draw a different net that can cover a box, and find its surface area.

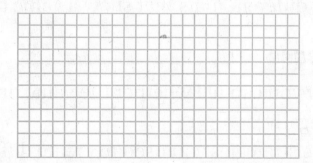

Think and Discuss

5. **Explain** how you can use a net to find surface area.

Holt Mathematics

11-1 Integers in Real-World Situations

Integers can be represented with two-color counters. A red counter represents −1, and a yellow counter represents 1. Each red and yellow pair is called a zero pair.

A zero pair has a value of zero. To find the value on an integer mat, remove zero pairs.

The value on this mat is −1.

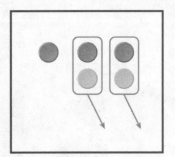

Find the value on each mat.

1.

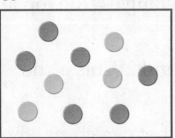

2.

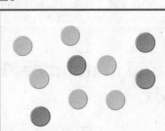

3.

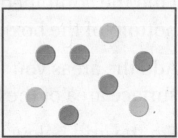

4. Create four different mats that show a value of −1.

Think and Discuss

5. Discuss how you can tell whether the value on a mat will be positive or negative.

6. Describe the strategies you used to create different mats that show a value of −1.

Holt Mathematics

11-2 Comparing and Ordering Integers

1. The completed table shows the average January temperatures in degrees Fahrenheit and degrees Celsius for some U.S. cities. Complete the other table by ordering the cities from warmest to coolest.

	°F	°C
Juneau, AK	24	−4
Phoenix, AZ	54	12
Atlanta, GA	41	5
Des Moines, IA	19	−7
Bismarck, ND	9	−13
Houston, TX	50	10
Boston, MA	29	−2
Kansas City, MO	26	−3

Source: Statistical Abstract of the United States

Warmest

	°F	°C

Coolest

Boston is colder than Houston because **29 < 50** in degrees Fahrenheit and **−2 < 10** in degrees Celsius.

Houston is warmer than Boston because **50 > 29** in degrees Fahrenheit and **10 > −2** in degrees Celsius.

2. Use inequality symbols to compare the Kansas City temperature in degrees Celsius with each of the other temperatures in degrees Celsius.

Think and Discuss

3. **Describe** your method for ordering the cities from warmest to coolest.

Holt Mathematics

11-3 The Coordinate Plane

On the *coordinate plane* below, the color of the first number in each ordered pair matches the color of the *x-axis,* and the color of the second number in each ordered pair matches the color of the *y-axis.*

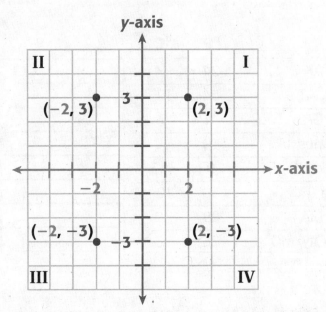

1. The four points graphed are labeled with their ordered pairs. How are these four ordered pairs alike? How are they different?

2. Plot the points $(3, 4)$, $(-3, 4)$, $(-3, -4)$, and $(3, -4)$ on the same coordinate plane.

Think and Discuss

3. **Describe** what each number in an ordered pair tells you.

Holt Mathematics

11-4 Adding Integers

You can use a thermometer to model addition of integers.

1. Suppose the temperature starts at −10°F and increases 30° during the day. Complete the addition statement to show the new temperature.

$$-10° + 30° = \underline{\hspace{1cm}}$$

2. Suppose the temperature starts at 20°F and drops 40° overnight. Complete the addition statement to show the new temperature.

$$20° + (-40°) = \underline{\hspace{1cm}}$$

3. Draw a thermometer and show $20° + (-10°)$. Find the sum.

Think and Discuss

4. **Describe** how to add a positive integer using a thermometer.
5. **Describe** how to add a negative integer using a thermometer.

Holt Mathematics

11-5 Subtracting Integers

You can use a number line to model subtracting integers.

To subtract 20 from 50, begin at the number being subtracted, 20, and count the number of units to the number 50.

The direction is **right**, so the difference is **positive**.

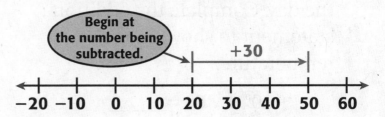

$50 - 20 = 30$

To subtract -20 from -60, begin at the number being subtracted, -20, and count the number of units to the number -60.

The direction is **left**, so the difference is **negative**.

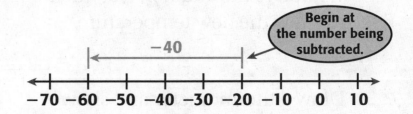

$-60 - (-20) = -40$

Use a number line to find each difference.

1. $12 - 10 =$ _____

2. $15 - 20 =$ _____

3. $-8 - 5 =$ _____

4. $-6 - (-4) =$ _____

5. $2 - (-4) =$ _____

6. $7 - (-2) =$ _____

Think and Discuss

7. Describe how to use a number line to model subtraction.

Holt Mathematics

11-6 Multiplying Integers

Complete each table.

1.

2 · 3 = 6	−2 · (3) = −6	2 · (−3) = −6	−2 · (−3) = 6
2 · 2 =	−2 · (2) =	2 · (−2) =	−2 · (−2) =
2 · 1 =	−2 · (1) =	2 · (−1) =	−2 · (−1) =

2.

3 · 3 =	−3 · (3) =	3 · (−3) =	−3 · (−3) =
3 · 2 =	−3 · (2) =	3 · (−2) =	−3 · (−2) =
3 · 1 =	−3 · (1) =	3 · (−1) =	−3 · (−1) =

3.

4 · 3 =	−4 · (3) =	4 · (−3) =	−4 · (−3) =
4 · 2 =	−4 · (2) =	4 · (−2) =	−4 · (−2) =
4 · 1 =	−4 · (1) =	4 · (−1) =	−4 · (−1) =

Think and Discuss

4. Describe the patterns you notice in each of the tables.

Holt Mathematics

11-7 Dividing Integers

For each multiplication statement, you can write two related division statements.

Multiplication statement	Division statements
$2 \cdot 3 = 6$	$6 \div 3 = 2$ and $6 \div 2 = 3$

Complete each table.

1.

Multiply	$4 \cdot (-3) =$	$-4 \cdot (-3) =$	$-4 \cdot 3 =$
Divide	$-12 \div 4 =$	$12 \div (-4) =$	$-12 \div (-4) =$
	$-12 \div (-3) =$	$12 \div (-3) =$	$-12 \div 3 =$

2.

Multiply	$2 \cdot (-5) =$	$-2 \cdot (-5) =$	$-2 \cdot 5 =$
Divide	$-10 \div 2 =$	$10 \div (-2) =$	$-10 \div (-2) =$
	$-10 \div (-5) =$	$10 \div (-5) =$	$-10 \div 5 =$

3.

Multiply	$8 \cdot (-3) =$	$-8 \cdot (-3) =$	$-8 \cdot 3 =$
Divide	$-24 \div 8 =$	$24 \div (-8) =$	$-24 \div (-8) =$
	$-24 \div (-3) =$	$24 \div (-3) =$	$-24 \div 3 =$

Think and Discuss

4. Describe what you think the sign rules are for dividing a positive integer by a negative integer, a negative integer by a positive integer, and a negative integer by a negative integer.

Holt Mathematics

EXPLORATION

11-8 Solving Integer Equations

You can use algebra tiles to model solving integer equations.

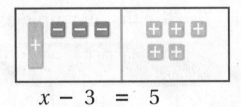

The equation $x - 3 = 5$ is modeled.

$$x - 3 = 5$$

To get x alone on one side, add three positive tiles to each side of the mat. This allows you to remove three zero pairs from the left side.

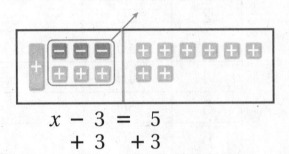

$$x - 3 = 5$$
$$+ 3 \quad + 3$$

The solution is 8.

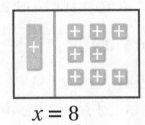

$$x = 8$$

Use algebra tiles to solve each equation.

1. $x + 5 = 9$ 2. $x - 6 = 2$ 3. $x + 4 = -1$

4. $6 = x - 7$ 5. $8 = x + 2$ 6. $3 = x - 9$

Think and Discuss

7. **Explain** how you know whether to add positive tiles or negative tiles to each side of the mat.

Holt Mathematics

EXPLORATION

11-9 Tables and Functions

A school has scheduled a trip for 210 students to a theme park. The school can rent up to six buses for $200 each. Each bus seats a maximum of 60 students. Tickets to the theme park cost $25 per student.

1. Use the first example as a guide to complete the table.

Group Number	Number of Students	Cost of Bus Rental and Tickets	Total Cost
1	10	$200 + 25 \cdot 10 = 200 + 250$	$450.00
2	20		
3	30		
4	40		
5	50		
6	60		

2. What numbers in the table remain constant?

3. What numbers in the table vary?

4. Use the cost of the bus ($200), the cost of each ticket ($25), and the number of students (x) to write an equation for the total cost (c). (*Hint:* Look at the middle column of the table to write the equation.)

Think and Discuss

5. **Explain** what makes the total cost of a group vary.

6. **Discuss** possible ways of reducing the total cost of taking 210 students to the theme park.

Holt Mathematics

11-10 Graphing Functions

To graph a *linear equation,* you can plot ordered pairs from a table of *x-* and *y*-values as points on a coordinate grid.

The table and graph at right model the function $y = x + 2$.

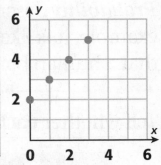

x	y
0	2
1	3
2	4
3	5

Graph each set of ordered pairs on the coordinate grid.

1. $y = x - 2$

x	y
2	0
3	1
4	2
5	3

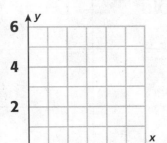

2. $y = x + 1$

x	y
1	2
2	3
3	4
4	5

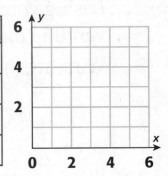

Use the points on each graph to complete each table.

3.

x	y

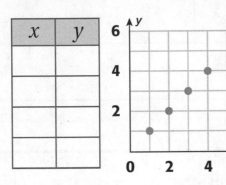

4.

x	y

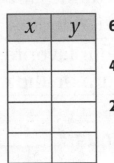
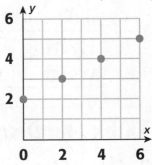

Think and Discuss

5. Explain how to use the points on the graph of a linear equation to write a table of ordered pairs.

Holt Mathematics

12-1 Introduction to Probability

Probability describes how likely it is that an event will occur. For example, it is likely that a dog will eat a treat, and it is unlikely that a human being will live 200 years.

Tell whether each event is likely or unlikely to happen.

	Event	Probability
1.	Having a blackout during a thunderstorm	
2.	Losing your money in an old vending machine	
3.	Winning the lottery	
4.	Finding a $100 bill on the street	
5.	Passing a test for which you studied very hard	
6.	Being in a traffic jam at 5:00 P.M. on Monday	
7.	Hearing your favorite song when you first turn on the radio	

Think and Discuss

8. Describe an event that is impossible.

9. Describe an event that is certain to happen.

Holt Mathematics

12-2 Experimental Probability

You can find the *experimental probability* of an event by dividing the number of times an event occurs by the total number of times the experiment is performed.

$$\text{probability} = \frac{\text{number of times an event occurs}}{\text{total number of trials}}$$

1. The data in the table show the number of free throws five players made in a season. Find the $\frac{\text{made}}{\text{attempts}}$ ratio for each player.

	Bo	Jack	Ali	Kim	José
Free Throws Made	30	32	15	36	24
Attempts	48	64	25	48	49
$\frac{\text{Made}}{\text{Attempts}}$					

2. Which player has the best chance of making a free throw?

3. Which player has the worst chance of making a free throw?

Think and Discuss

4. **Discuss** how you determined the answers for numbers 2 and 3.

5. **Explain** how to write each $\frac{\text{made}}{\text{attempts}}$ ratio as a percent.

Holt Mathematics

12-3 Counting Methods and Sample Spaces

A teacher made a quiz with three true or false questions. You can make an organized list to determine all of the answer possibilities for the three questions.

1. Complete the tree diagram to list all possible answer outcomes.

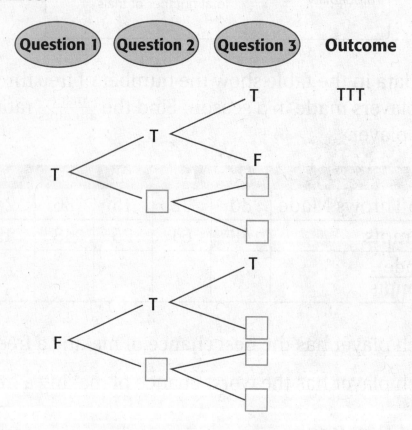

| Question 1 | Question 2 | Question 3 | Outcome |

2. How many outcomes are possible?

Think and Discuss

3. **Discuss** your method for organizing the list of outcomes.

4. **Explain** how you could determine the number of outcomes possible with four questions.

Holt Mathematics

12-4 Theoretical Probability

When you flip a fair coin, the *theoretical probability* of getting tails is 50% and the theoretical probability of getting heads is 50%. These two outcomes are equally likely to occur.

Determine whether the outcomes in each experiment are all equally likely to occur.

	Equally Likely	Not Equally Likely
1.		
2.		
3.		

Think and Discuss

4. Explain how you determined whether the outcomes in numbers **1–3** were equally likely or not.

Holt Mathematics

12-5 Compound Events

A *compound event* consists of two or more single events.

A bat had two babies. The birth of each baby bat is a single event. The birth of both bats is a compound event.

First bat born

	Female	Male
Female	FF	MF
Male	FM	MM

Second bat born

1. List all possible outcomes for two offspring.

2. Divide 1 (FF) by the total number of outcomes to find the probability that both bats are female.

3. Divide 1 (MM) by the total number of outcomes to find the probability that both bats are male.

4. Divide 2 (MF and FM) by the total number of outcomes to find the probability that one bat is female and the other is male.

Think and Discuss _____

5. **Explain** how to find the probability that three males would be born among four births.

Holt Mathematics

12-6 Making Predictions

You can use probabilities to make *predictions*. For example, the probability of rolling a 1 on a number cube is $\frac{1}{6}$. If you roll the cube 12 times, how many times do you predict you will roll a 1?

Each time you roll the cube, the probability of rolling a 1 is $\frac{1}{6}$.

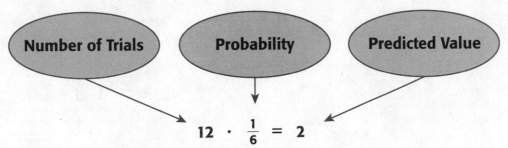

$$12 \cdot \frac{1}{6} = 2$$

So you expect to roll a 1 twice when you roll a cube 12 times.

Use the model as a guide to make each prediction.

	Event	Probability	Number of Trials	Predicted Value
1.	Getting heads when flipping a coin	$\frac{1}{2}$	Flip the coin 20 times.	
2.	Spinning a 2 on a spinner divided into 4 equal sections	$\frac{1}{4}$	Spin the spinner 24 times.	

Think and Discuss

3. **Explain** how to find a predicted value.

4. **Discuss** whether it is certain that you will roll a 1 twice when you roll a number cube 12 times.

Holt Mathematics

Name _____ Date _____ Class _____

Problem Solving
Comparing and Ordering Whole Numbers

Use the tables below to answer each question.

Most Populated Countries	
Brazil	174,468,575
China	1,273,111,290
India	1,029,991,145
Indonesia	228,437,870
United States	278,058,881

Largest Countries (square mi)	
Brazil	3,265,059
Canada	3,849,646
China	3,705,408
Russia	6,592,812
United States	3,539,224

1. Which country has the greatest population?

_____

2. Which countries have more than one billion people?

_____

3. Which country is the largest in the world?

_____

4. Which country's area is closest to 4,000,000 square miles?

_____

5. What is the error in the following statement? Canada is larger than the United States, but smaller than China.

_____

_____

6. Based on population and size, which country do you think is more crowded, Brazil or the United States? Explain.

_____

_____

7. Which country has a population less than two hundred million?

  A China   **C** Brazil

  B Indonesia   **D** India

9. Which list shows the countries in order by population from greatest to least?

  A China, United States, India, Indonesia, Brazil

  B China, India, Indonesia, Brazil, United States,

  C China, India, Indonesia, United States, Brazil

  D China, India, United States, Indonesia, Brazil

8. Which countries have populations greater than the United States?

  F China and Brazil

  G China and India

  H India and Indonesia

  J Indonesia and China

10. Which list shows the countries in order by size from smallest to largest?

  F Brazil, United States, China, Canada, Russia

  G Brazil, United States, Canada, China, Russia

  H Brazil, United States, Canada, Russia, China

  J Brazil, United States, Russia, China, Canada

Holt Mathematics

Name _____ Date _____ Class _____

Problem Solving
Estimating with Whole Numbers

Use the table below to answer each question.

Facts About the World's Oceans

Ocean	Area (square mi)	Greatest Depth (ft)
Arctic	5,108,132	18,456
Atlantic	33,424,006	30,246
Indian	28,351,484	24,460
Pacific	64,185,629	35,837

1. If the depths of all the oceans were rounded to the nearest ten thousand, which two oceans would have the same depth?

   _____

2. In 1960, scientists observed sea creatures living as far down as thirty thousand feet. In which ocean(s) could these creatures have lived?

   _____

3. If you wanted to compare the depths of the Pacific Ocean and the Atlantic Ocean, which place value would you use to estimate?

   _____

4. The oceans cover about three-fourths of Earth's surface. Estimate the total area of all the oceans combined by rounding to the nearest million.

   _____

Choose the letter for the best answer.

5. There are 5,280 feet in a mile. About how many miles deep is the deepest point in the Pacific Ocean?

   A about 0.7 mile   C about 70 miles

   B about 7 miles   D about 700 miles

6. Rounding to the greatest place value, about how much larger is the Indian Ocean than the Arctic Ocean?

   F about 5 million sq. mi

   G about 10 million sq. mi

   H about 15 million sq. mi

   J about 25 million sq. mi

7. The Atlantic Ocean is about 40 times larger than the world's largest island, Greenland. Use this information to estimate the area of Greenland.

   A about 800,000 sq. mi

   B about 8,000,000 sq. mi

   C about 80,000,000 sq. mi

   D about 1,200,000,000 sq. mi

8. About how much larger would the Pacific Ocean have to be to have more area than the other three oceans combined?

   F about 2 hundred sq. mi

   G about 2 thousand sq. mi

   H about 2 million sq. mi

   J about 20 million sq. mi

Holt Mathematics

Problem Solving
LESSON 1-3 Exponents

1. The Sun is the center of our solar system. The Sun is the star closest to our planet. The surface temperature of the Sun is close to 10,000°F. Write 10,000 using exponents.

2. Patty Berg has won 4^2 major women's titles in golf. Write 4^2 in standard form.

3. William has 3^3 baseball cards and 4^3 football cards. Write the number of baseball cards and footballs cards that William has.

4. Michelle recorded the number of miles she ran each day last year. She used the following expression to represent the total number of miles: $3 \times 3 \times 3 \times 3 \times 3 \times 3 \times 3$. Write this expression using exponents. How many miles did Michelle run last year?

Choose the letter for the best answer.

5. In Tyrone's science class he is studying cells. Cell A divides every 30 minutes. If Tyrone starts with two cells, how many cells will he have in 3 hours?

   A 6 cells

   B 32 cells

   C 128 cells

   D 512 cells

6. Tanisha's soccer team has a phone tree in case a soccer game is postponed or cancelled. The coach calls 2 families. Then each family calls 2 other families. How many families will be notified during the 4th round of calls?

   F 2 families

   G 4 families

   H 8 families

   J 16 families

7. The Akashi-Kaiko Bridge is the longest suspension bridge in the world. It is located in Kobe-Naruto, Japan and was completed in 1998. It is about 3^8 feet long. Write the approximate length of the Akashi-Kaiko Bridge in standard form.

   A 6,561 feet

   B 2,187 feet

   C 512 feet

   D 24 feet

8. The Strahov Stadium is the largest sports stadium in the world. It is located in Prague, Czech Republic. Its capacity is about 12^5 people. Write the capacity of the Strahov Stadium in standard form.

   F 60 people

   G 144 people

   H 20,736 people

   J 248,832 people

Holt Mathematics

LESSON 1-4

Problem Solving
Order of Operations

Evaluate each expression to complete the table.

Mammals with the Longest Tails

	Mammal	Expression	Tail Length
1.	Asian elephant	$2 + 3^2 \times 7 - (10 - 4)$	
2.	Leopard	$5 \times 6 + 5^2$	
3.	African elephant	$6 \times (72 \div 8) - 3$	
4.	African buffalo	$51 + 6^2 \div 9 - 12$	
5.	Giraffe	$4^3 - 3 \times 7$	
6.	Red kangaroo	$11 + 48 \div 6 \times 4$	

Choose the letter for the best answer.

7. Adam and his two brothers went to the zoo. Each ticket to enter the zoo costs $7. Adam bought two bags of peanuts for $4 each, and one of his brothers bought a lion poster for $12. Which expression shows how much money they spent at the zoo in all?

   A $7 + 4 + 12$

   B $7 \times 3 + 4 + 12$

   C $7 \times 3 + 4 \times 2 + 12$

   D $(7 \times 3) + (4 \times 12)$

8. An elephant eats about 500 pounds of grass and leaves every day. There are 2 Africa elephants and 3 Asian elephants living in the City Zoo. How many pounds of grass and leaves do the zookeepers need to order each week to feed all the elephants?

   F 2,500 pounds

   G 17,500 pounds

   H 3,000 pounds

   J 21,000 pounds

9. The average giraffe is 18 feet tall. Which of these expressions shows the height of a giraffe?

   A $4^2 - 2$

   B $3 \times 12 \div 4 + 2$

   C $3^3 \div 9 \times 6$

   D $20 \div 5 + 5 - 6$

10. Some kangaroos can cover 30 feet in a single jump! If a kangaroo could jump like that 150 times in a row, how much farther would it need to go to cover a mile? (1 mile = 5,280 feet)

   F 780 feet   **H** 176 feet

   G 26 feet   **J** 5,100 feet

Holt Mathematics

Name _____ Date _____ Class _____

Problem Solving
LESSON 1-5 *Mental Math*

The bar graph below shows the average amounts of water used during some daily activities. Use the bar graph and mental math to answer the questions.

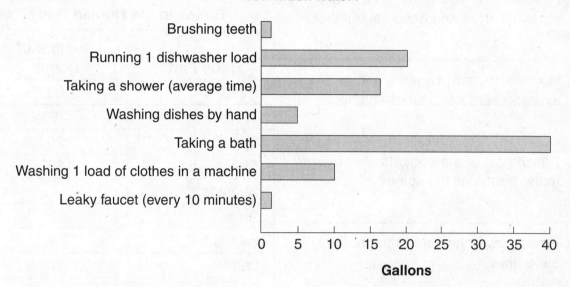

How Much Water?

1. Most people brush their teeth three times a day. How much water do they use for this activity every week?

   _____

2. How much water is wasted in a day by a leaky faucet?

   _____

3. The average American uses 124 gallons of water a day. Name a combination of activities listed in the table that would equal that daily total.

   _____

   _____

Choose the letter for the best answer.

4. Kenya used 24 gallons of water doing three of the activities listed in the table once. Which activities did she do?

   A taking a bath, brushing teeth, washing dishes by hand

   B taking a bath, brushing teeth, running 1 dishwasher load

   C taking a shower, brushing teeth, washing dishes by hand

   D taking a shower, brushing teeth, running 1 dishwasher load

5. If you wash two loads of dishes by hand instead of using a dishwasher, how much water do you save?

   F 30 gallons   **G** 15 gallons   **H** 10 gallons   **J** 1 gallon

Holt Mathematics

Problem Solving

Choose the Method of Computation

Use the table below to answer questions 1–6. For each question, write the method of computation you should use to solve it. Then write the solution.

1. How many bones are in an average person's arms and hands altogether?

   _____

2. How many more bones are in an average person's head than chest?

   _____

3. Which part of the body has twice as many bones as the spine?

   _____

4. How many bones are in the body altogether?

   _____

Bones in the Human Body

Body Part	Number of Bones
Head	28
Throat	1
Spine	26
Chest	25
Shoulders	4
Arms	6
Hands	54
Legs	10
Feet	52

5. A newborn baby has 350 bones. How many more bones does a newborn baby have than an adult?

   _____

   _____

6. How many bones are in each of an average person's feet, hands, legs, and arms?

   _____

   _____

Choose the letter for the best answer.

7. The body's longest bones—thighbones and shinbones—are in the legs. The average thighbone is about 20 inches long, and the average shinbone is about 17 inches long. What is the total length of those four bones?

   A paper and pencil; 74 inches

   B paper and pencil; 37 inches

   C mental math; 20 inches

   D calculator; 17 inches

8. The body has 650 muscles. Seventeen of those muscles are used to smile and 42 muscles are used to frown. How many more muscles are used to frown than to smile?

   F mental math; 35 muscles

   G mental math; 25 muscles

   H paper and pencil; 608 muscles

   J calculator; 633 muscles

Holt Mathematics

Name _____ Date _____ Class _____

Problem Solving
Patterns and Sequences

1. A giant bamboo plant was 5 inches tall on Monday, 23 inches tall on Tuesday, 41 inches tall on Wednesday, and 59 inches tall on Thursday. Describe the pattern. If the pattern continues, how tall will the giant bamboo plant be on Friday, Saturday, and Sunday?

   _____

   _____

   _____

2. A scientist was studying a cell. After the second hour there were two cells. After the third hour there were four cells. After the fourth hour there were eight cells. Describe the pattern. If the pattern continues, how many cells will there be after the fifth, sixth, and seventh hour?

   _____

   _____

   _____

Choose the letter for the best answer.

3. The first place prize for a sweepstakes is $8,000. The third place prize is $2,000. The fourth place prize is $1,000. The fifth place prize is $500. What is the second place prize?

   A $7,000   **C** $4,000

   B $6,000   **D** $3,000

4. The temperature was 59°F at 3:00 A.M., 62°F at 5:00 A.M., and 65°F at 7:00 A.M. If the pattern continues, what will the temperature be at 9:00 A.M., 11:00 A.M., and 1:00 P.M.?

   F 66°F at 9:00 A.M., 67°F at 11:00 A.M., 68°F at 1:00 P.M.

   G 68°F at 9:00 A.M., 70°F at 11:00 A.M., 72°F at 1:00 P.M.

   H 68°F at 9:00 A.M., 71°F at 11:00 A.M., 74°F at 1:00 P.M.

   J 70°F at 9:00 A.M., 75°F at 11:00 A.M., 80°F at 1:00 P.M.

Holt Mathematics

Name _____ Date _____ Class _____

Problem Solving
Variables and Expressions

Write the correct answer.

1. To cook 4 cups of rice, you use 8 cups of water. To cook 10 cups of rice, you use 20 cups of water. Write an expression to show how many cups of water you should use if you want to cook c cups of rice. How many cups of water should you use to cook 5 cups of rice?

2. Sue earns the same amount of money for each hour that she tutors students in math. In 3 hours, she earns $27. In 8 hours, she earns $72. Write an expression to show how much money Sue earns working h hours. At this rate, how much money will Sue earn if she works 12 hours?

3. Bees are one of the fastest insects on Earth. They can fly 22 miles in 2 hours, and 55 miles in 5 hours. Write an expression to show how many miles a bee can fly in h hours. If a bee flies 4 hours at this speed, how many miles will it travel?

4. A friend asks you to think of a number, triple it, and then subtract 2. Write an algebraic expression using the variable x to describe your friend's directions. Then find the value of the expression if the number you think of is 5.

Circle the letter of the correct answer.

5. The ruble is the currency in Russia. In 2005, 1 United States dollar was worth 28 rubles. How many rubles were equivalent to 10 United States dollars?

   A 28

   B 38

   C 280

   D 2,800

6. The peso is the currency in Mexico. In 2005, 1 United States dollar was worth 10 pesos. How many pesos were equivalent to 5 United States dollars?

   F 1

   G 10

   H 15

   J 50

Holt Mathematics

Problem Solving

LESSON 2-2 *Translate Between Words and Math*

Write the correct answer.

21. Holly bought 10 comic books. She gave a few of them to Kyle. Let *c* represent the number of comic books she gave to Kyle. Write an expression for the number of comic books Holly has left.

2. Last week, Peter worked 40 hours for $15 an hour. Write a numerical expression for the total amount Peter earned last week. Write an algebraic expression to show how much Peter earns in *h* hours at that rate.

3. The temperature dropped 5°F, and then it went up 3°F. Let *t* represent the beginning temperature. Write an expression to show the ending temperature.

4. Teri baked 48 cookies and divided them evenly into bags. Let *n* represent the number of cookies Teri put in each bag. Write an expression for the number of bags she filled.

Circle the letter of the correct answer.

5. Marisa purchased canned soft drinks for a family reunion. She purchased 1 case of 24 cans and several packages containing 6 cans each. If *p* represents the number of 6-can packages she purchased, which of the following expressions represents the total number of cans Marisa purchased for the reunion?

A $24 + 6p$

B $24 - 6p$

C $6 + 24p$

D $6 - 24p$

6. Becky has the addresses of many people listed in her e-mail address book. She forwarded a copy of an article to all but 5 of those people. If *a* represents the number of addresses, which of the following expressions represents how many people she sent the article to?

F $a + 5$

G $5a$

H $a - 5$

J $a \div 5$

7. Mei bought several CDs for $12 each. Which of the following expressions could you use to find the total amount she spent on the CDs?

A $12 + x$

B $12 - x$

C $12x$

D $12 \div x$

8. Tony bought 2 packs of 50 plates and 1 pack of 30 plates. Which of the following expressions could you use to find the total number of plates Tony bought?

F $2 + 50 + 30$

G $(2 \cdot 50) + 30$

H $(2 \cdot 30) + 50$

J $2(30 + 50)$

Holt Mathematics

Name _____ Date _____ Class _____

Problem Solving
Translating Between Tables and Expressions

Use the table to write an expression for the missing value.
Then use your expression to answer the questions.

1. How many cars are produced on average each year?

_____

2. How many cars will be produced in 6 years?

_____

3. After how many years will there be an average production of 3,750 cars?

_____

Cars Produced By Company X

Number of Years	Average Number of Cars Produced
2	2,500
5	6,250
7	8,750
10	12,500
12	15,000
14	17,500
n	

Circle the letter of the correct answer.
Company Y produces twice as many cars as Company X.

4. How many cars does Company Y produce on average in 8 years?

A 1,250
B 10,000
C 11,250
D 20,000

5. How many more cars on average does Company Y produce in 4 years than Company X?

F 2,500
G 5,000
H 6,125
J 7,500

6. Which company produces an average of 11,250 cars in 9 years?

A Company X
B Company Y
C both companies
D neither company

7. How many cars are produced on average by both companies in 20 years?

F 3,750
G 12,500
H 25,000
J 37,500

Holt Mathematics

Problem Solving

LESSON 2-4

Equations and Their Solutions

Use the table to write and solve an equation to answer each question. Then use your answers to complete the table.

1. A hippopotamus can stay underwater 3 times as long as a sea otter can. How long can a sea otter stay underwater?

   _____

   _____

2. A seal can stay underwater 10 minutes longer than a muskrat can. How long can a muskrat stay underwater?

   _____

   _____

3. A sperm whale can stay underwater 7 times longer than a sea cow can. How long can a sperm whale stay underwater?

   _____

   _____

How Many Minutes Can Mammals Stay Underwater?	
Hippopotamus	15
Human	
Muskrat	
Platypus	10
Polar bear	
Sea cow	16
Sea otter	
Seal	22
Sperm whale	

Circle the letter of the correct answer.

4. The difference between the time a platypus and a polar bear can stay underwater is 8 minutes. How long can a polar bear stay underwater?

   A 1 minute

   B 2 minutes

   C 3 minutes

   D 5 minutes

5. When you divide the amount of time any of the animals in the table can stay underwater by itself, the answer is always the amount of time the average human can stay underwater. How long can the average human stay underwater?

   F 6 minutes

   G 4 minutes

   H 2 minutes

   J 1 minute

Holt Mathematics

Name _____ Date _____ Class _____

Problem Solving
Addition Equations

Use the bar graph and addition equations to answer the questions.

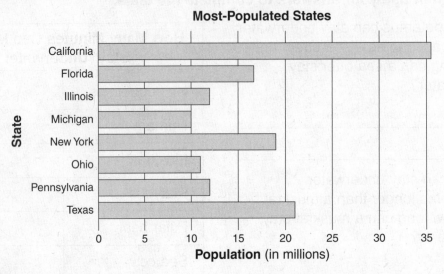

Most-Populated States

1. How many more people live in California than in New York?

   _____

   _____

2. How many more people live in Ohio than in Michigan?

   _____

   _____

3. How many more people live in Florida than in Illinois?

   _____

   _____

4. How many more people live in Texas than in Pennsylvania?

   _____

   _____

Circle the letter of the correct answer.

5. Which two states' populations are used in the equation $12 + x = 12$?

   A Pennsylvania and Texas

   B Ohio and Florida

   C Michigan and Illinois

   D Illinois and Pennsylvania

6. What is the value of x in the equation in Exercise 5?

   F 0   H 12

   G 1   J 24

7. In 2003, the total population of the United States was 292 million. How many of those people did not live in one of the states shown on the graph?

   A 416 million   C 154 million

   B 73 million   D 292 million

8. The combined population of Ohio and one other state is the same as the population of Texas. What is that state?

   F California

   G Florida

   H Michigan

   J Pennsylvania

Holt Mathematics

Problem Solving
Subtraction Equations

Write and solve subtraction equations to answer the questions.

1. Dr. Felix Hoffman invented aspirin in 1899. That was 29 years before Alexander Fleming invented penicillin. When was penicillin invented?

_____

_____

2. Kimberly was born on February 2. That is 10 days earlier than Kent's birthday. When is Kent's birthday?

_____

_____

3. Kansas and North Dakota are the top wheat-producing states. In 2000, North Dakota produced 314 million bushels of wheat, which was 34 million bushels less than Kansas produced. How much wheat did Kansas farmers grow in 2000?

_____

_____

4. Scientists assign every element an atomic number, which is the number of protons in the nucleus of that element. The atomic number of silver is 47, which is 32 less than the atomic number of gold. How many protons are in the nucleus of gold?

_____

_____

Circle the letter of the correct answer.

5. The spine-tailed swift and the frigate bird are the two fastest birds on earth. A frigate bird can fly 95 miles per hour, which is 11 miles per hour slower than a spine-tailed swift. How fast can a spine-tailed swift fly?

   A 84 miles per hour

   B 101 miles per hour

   C 106 miles per hour

   D 116 miles per hour

6. The Green Bay Packers and the Kansas City Chiefs played in the first Super Bowl in 1967. The Chiefs lost by 25 points, with a final score of 10. How many points did the Packers score in the first Super Bowl?

   F 35

   G 25

   H 15

   J 0

7. The Rocky Mountains extend 3,750 miles across North America. That is 750 miles shorter than the Andes Mountains in South America. How long are the Andes Mountains?

   A 3,000 miles   C 180 miles

   B 5 miles   D 4,500 miles

8. When the United States took its first census in 1790, only 4 million people lived here. That was 288 million fewer people than the population in 2003. What was the population of the United States in 2003?

   F 292 million   H 69 million

   G 284 million   J 1,108 million

Holt Mathematics

Name _____ Date _____ Class _____

Problem Solving
Multiplication Equations

Write and solve a multiplication equation to answer each question.

1. In 1975, a person earning minimum wage made $80 for a 40-hour work week. What was the minimum wage per hour in 1975?

   _____

   _____

2. If an ostrich could maintain its maximum speed for 5 hours, it could run 225 miles. How fast can an ostrich run?

   _____

   _____

3. About 2,000,000 people live in Paris, the capital of France. That is 80 times larger than the population of Paris, Texas. How many people live in Paris, Texas?

   _____

   _____

4. The average person in China goes to the movies 12 times per year. That is 3 times more than the average American goes to the movies. How many times per year does the average American go to the movies?

   _____

   _____

Circle the letter of the correct answer.

5. Recycling just 1 ton of paper saves 17 trees! If a city recycled enough paper to save 136 trees, how many tons of paper did it recycle?

   A 7 tons

   B 8 tons

   C 9 tons

   D 119 tons

6. Seaweed found along the coast of California, called giant kelp, grows up to 18 inches per day. If a giant kelp plant has grown 162 inches at this rate, for how many days has it been growing?

   F 180 days   **H** 9 days

   G 144 days   **J** 8 days

7. The distance between Atlanta, Georgia, and Denver, Colorado, is 1,398 miles. That is twice the distance between Atlanta and Detroit, Michigan. How many miles would you have to drive to get from Atlanta to Detroit?

   A 2,796 miles

   B 349.5 miles

   C 699 miles

   D 1,400 miles

8. Jupiter has 2 times more moons than Neptune has, and 8 times more moons than Mars has. Jupiter has 16 moons. How many moons do Neptune and Mars each have?

   F 8 moons, 2 moons

   G 2 moons, 8 moons

   H 128 moons, 32 moons

   J 32 moons, 128 moons

Holt Mathematics

Name _____ Date _____ Class _____

Problem Solving
Division Equations

Use the table to write and solve a division equation to answer each question.

1. How many total people signed up to play soccer in Bakersville this year?

   _____

2. How many people signed up to play lacrosse this year?

   _____

3. What was the total number of people who signed up to play baseball this year?

   _____

4. Which two sports in the league have the same number of people signed up to play this year? How many people are signed up to play each of those sports?

   _____

Bakersville Sports League

Sport	Number of Teams	Players on Each Team
Baseball	7	20
Soccer	11	15
Football	8	24
Volleyball	12	9
Lacrosse	6	17
Basketball	10	10
Tennis	18	6

Circle the letter of the correct answer.

5. Which sport has a higher total number of players, football or tennis? How many more players?

   A football; 10 players

   B tennis; 144 players

   C football; 84 players

   D tennis; 18 players

6. Only one sport this year has the same number of players on each team as its number of teams. Which sport is that?

   F basketball

   G football

   H soccer

   J tennis

Holt Mathematics

Problem Solving

Representing, Comparing, and Ordering Decimals

Use the table to answer the questions.

1. What is the heaviest marine
   mammal on Earth?

   _____

2. Which mammal in the table has
   the shortest length?

   _____

3. Which mammal in the table is longer
   than a humpback whale, but shorter
   than a sperm whale?

   _____

Largest Marine Mammals

Mammal	Length (ft)	Weight (T)
Blue whale	110.0	127.95
Fin whale	82.0	44.29
Gray whale	46.0	32.18
Humpback whale	49.2	26.08
Right whale	57.4	39.37
Sperm whale	59.0	35.43

Circle the letter of the correct answer.

4. Which mammal measures forty-nine
   and two tenths feet long?

   A blue whale

   B gray whale

   C sperm whale

   D humpback whale

5. Which mammal weighs thirty-five and
   forty-three hundredths tons?

   F right whale

   G sperm whale

   H gray whale

   J fin whale

6. Which of the following lists shows
   mammals in order from the least
   weight to the greatest weight?

   A sperm whale, right whale,
   fin whale, gray whale

   B fin whale, sperm whale, gray
   whale, blue whale

   C fin whale, right whale, sperm
   whale, gray whale

   D gray whale, sperm whale,
   right whale, fin whale

7. Which of the following lists shows
   mammals in order from the greatest
   length to the least length?

   F sperm whale, right whale,
   humpback whale, gray whale

   G gray whale, humpback whale,
   right whale, sperm whale

   H right whale, sperm whale, gray
   whale, humpback whale

   J humpback whale, gray whale,
   sperm whale, right whale

Holt Mathematics

LESSON
3-2
Problem Solving
Estimating Decimals

Write the correct answer.

1. Men in Iceland have the highest average life expectancy in the world—76.8 years. The average life expectancy for a man in the United States is 73.1 years. About how much higher is a man's average life expectancy in Iceland? Round your answer to the nearest whole year.

2. The average life expectancy for a woman in the United States is 79.1 years. Women in Japan have the highest average life expectancy—3.4 years higher than the United States. Estimate the average life expectancy of women in Japan. Round your answer to the nearest whole year.

3. There are about 1.6093 kilometers in one mile. There are 26.2 miles in a marathon race. About how many kilometers are there in a marathon race? Round your answer to the nearest tenths.

4. At top speed, a hornet can fly 13.39 miles per hour. About how many hours would it take a hornet to fly 65 miles? Round your answer to the nearest whole number.

Circle the letter of the correct answer.

5. The average male human brain weighs 49.7 ounces. The average female human brain weighs 44.6 ounces. What is the difference in their weights?

   A about 95 ounces

   B about 7 ounces

   C about 5 ounces

   D about 3 ounces

6. An official hockey puck is 2.54 centimeters thick. About how thick are two hockey pucks when one is placed on top of the other?

   F about 4 centimeters

   G about 4.2 centimeters

   H about 5 centimeters

   J about 5.2 centimeters

7. Lydia earned $9.75 per hour as a lifeguard last summer. She worked 25 hours a week. About how much did she earn in 8 weeks?

   A about $250.00

   B about $2,000.00

   C about $2,500.00

   D about $200.00

8. Brent mixed 4.5 gallons of blue paint with 1.7 gallons of white paint and 2.4 gallons of red paint to make a light purple paint. About how many gallons of purple paint did he make?

   F about 9 gallons

   G about 8 gallons

   H about 10 gallons

   J about 7 gallons

Holt Mathematics

LESSON 3-3 Problem Solving
Adding and Subtracting Decimals

Use the table to answer the questions.

Busiest Ports in the United States

Port	Imports Per Year (millions of tons)	Exports Per Year (millions of tons)
South Louisiana, LA	30.6	57.42
Houston, TX	75.12	33.43
New York, NY & NJ	53.52	8.03
New Orleans, LA	26.38	21.73
Corpus Christi, TX	52.6	7.64

1. How many more tons of imports than exports does the Port of New Orleans handle each year?

_____

2. How many tons of imports and exports are shipped through the port of Houston, Texas, each year in all?

_____

Circle the letter of the correct answer.

3. Which port ships 0.39 more tons of exports each year than the port at Corpus Christi, Texas?

   A Houston

   B NY & NJ

   C New Orleans

   D South Louisiana

4. What is the difference between the imports and exports shipped in and out of Corpus Christi's port each year?

   F 45.04 million tons

   G 44.94 million tons

   H 44.96 million tons

   J 44.06 million tons

5. What is the total amount of imports shipped into the nation's 5 busiest ports each year?

   A 238.22 million tons

   B 366.47 million tons

   C 128.25 million tons

   D 109.97 million tons

6. What is the total amount of exports shipped out of the nation's 5 busiest ports each year?

   F 366.47 million tons

   G 128.25 million tons

   H 109.97 million tons

   J 238.22 million tons

Holt Mathematics

Name _____ Date _____ Class _____

Write the correct answer.

1. The closest comet to approach Earth was called Lexell. On July 1, 1770, Lexell was observed about 874,200 miles from Earth's surface. Write this distance in scientific notation.

2. Scientists estimate that it would take $1.4 \cdot 10^{10}$ years for light from the edge of our universe to reach Earth. How many years is that written in standard form?

3. In the United States, about 229,000,000 people speak English. About 18,000,000 people speak English in Canada. Write in scientific notation the total number of English speaking people in the United States and Canada.

4. South Africa is the top gold-producing country in the world. Each year it produces $4.688 \cdot 10^8$ tons of gold! Written in standard form, how many tons of gold does South African produce each year?

Circle the letter of the correct answer.

5. About $3.012 \cdot 10^6$ people visit Yellowstone National Park each year. What is that figure written in standard form?
   A 30,120,000 people
   B 3,012,000 people
   C 301,200 people
   D 30,120 people

6. In 2000, farmers in Iowa grew 1,740,000 bushels of corn. What is this amount written in scientific notation?
   F $1.7 \cdot 10^5$
   G $1.74 \cdot 10^5$
   H $1.74 \cdot 10^6$
   J $1.74 \cdot 10^7$

7. The temperature at the core of the Sun reaches 27,720,000°F. What is this temperature written in scientific notation?
   A $2.7 \cdot 10^7$
   B $2.72 \cdot 10^7$
   C $2.772 \cdot 10^6$
   D $2.772 \cdot 10^7$

8. Your body is constantly producing red blood cells—about $1.73 \cdot 10^{11}$ cells a day. How many blood cells is that written in standard form?
   F 173,000,000 cells
   G 17,300,000,000 cells
   H 173,000,000,000 cells
   J 1,730,000,000,000 cells

Holt Mathematics

Problem Solving

Multiplying Decimals

Use the table to answer the questions.

1. At the minimum wage, how much did a person earn for a 40-hour workweek in 1950?

   _____

2. At the minimum wage, how much did a person earn for working 25 hours in 1970?

   _____

3. If you had a minimum-wage job in 1990, and worked 15 hours a week, how much would you have earned each week?

   _____

United States Minimum Wage

Year	Hourly Rate
1940	$0.30
1950	$0.75
1960	$1.00
1970	$1.60
1980	$3.10
1990	$3.80
2000	$5.15

4. About how many times higher was the minimum wage in 1960 than in 1940?

   _____

Circle the letter for the correct answer.

5. Ted's grandfather had a minimum-wage job in 1940. He worked 40 hours a week for the entire year. How much did Ted's grandfather earn in 1940?

   A $12.00

   B $624.00

   C $642.00

   D $6,240.00

6. Marci's mother had a minimum-wage job in 1980. She worked 12 hours a week. How much did Marci's mother earn each week?

   F $3.72

   G $37.00

   H $37.10

   J $37.20

7. Having one dollar in 1960 is equivalent to having $5.82 today. If you worked 40 hours a week in 1960 at minimum wage, how much would your weekly earnings be worth today?

   A $40.00

   B $5.82

   C $232.80

   D $2,328.00

8. In 2000, Cindy had a part-time job at a florist, where she earned minimum wage. She worked 18 hours each week for the whole year. How much did she earn from this job in 2000?

   F $927.00

   G $4,820.40

   H $10,712.00

   J $2,142.40

Holt Mathematics

Problem Solving

LESSON 3-6

Dividing Decimals by Whole Numbers

Write the correct answer.

1. Four friends had lunch together. The total bill for lunch came to $33.40, including tip. If they shared the bill equally, how much did they each pay?

2. There are 7.2 milligrams of iron in a dozen eggs. Because there are 12 eggs in a dozen, how many milligrams of iron are in 1 egg?

3. Kyle bought a sheet of lumber 8.7 feet long to build fence rails. He cut the strip into 3 equal pieces. How long is each piece?

4. An albatross has a wingspan greater than the length of a car—3.7 meters! Wingspan is the length from the tip of one wing to the tip of the other wing. What is the length of each albatross wing (assuming wing goes from center of body)?

Circle the letter of the correct answer.

5. The City Zoo feeds its three giant pandas 181.5 pounds of bamboo shoots every day. Each panda is fed the same amount of bamboo. How many pounds of bamboo does each panda eat every day?

   A 6.05 pounds

   B 60.5 pounds

   C 61.5 pounds

   D 605 pounds

6. Emma bought 22.5 yards of cloth to make curtains for two windows in her apartment. She used the same amount of cloth on each window. How much cloth did she use to make each set of curtains?

   F 1.125 yards

   G 10.25 yards

   H 11.25 yards

   J 11.52 yards

7. Aerobics classes cost $153.86 for 14 sessions. What is the fee for one session?

   A $10.99

   B $1.99

   C about $25.00

   D about $20.00

8. An entire apple pie has 36.8 grams of saturated fat. If the pie is cut into 8 slices, how many grams of saturated fat are in each slice?

   F 4.1 grams

   G 0.46 grams

   H 4.6 grams

   J 4.11 grams

Holt Mathematics

LESSON 3-7
Problem Solving
Dividing by Decimals

Write the correct answer.

1. Jamal spent $6.75 on wire to build a rabbit hutch. Wire costs $0.45 per foot. How many feet of wire did Jamal buy?

_____

2. Peter drove 195.3 miles in 3.5 hours. On average, how many miles per hour did he drive?

_____

3. Lisa's family drove 830.76 miles to visit her grandparents. Lisa calculated that they used 30.1 gallons of gas. How many miles per gallon did the car average?

_____

4. A chef bought 84.5 pounds of ground beef. He uses 0.5 pound of ground beef for each hamburger. How many hamburgers can he make?

_____

Circle the letter of the correct answer.

5. Mark earned $276.36 for working 23.5 hours last week. He earned the same amount of money for each hour that he worked. What is Mark's hourly rate of pay?

A $1.17

B $10.76

C $11.76

D $117.60

6. Alicia wants to cover a section of her wall that is 2 feet wide and 12 feet long with mirrors. Each mirror tile is 2 feet wide and 1.5 feet long. How many mirror tiles does she need to cover that section?

F 4 tiles

G 6 tiles

H 8 tiles

J 12 tiles

7. John ran the city marathon in 196.5 minutes. The marathon is 26.2 miles long. On average, how many miles per hour did John run the race?

A 7 miles per hour

B 6.2 miles per hour

C 8 miles per hour

D 8.5 miles per hour

8. Shaneeka is saving $5.75 of her allowance each week to buy a new camera that costs $51.75. How many weeks will she have to save to have enough money to buy it?

F 9 weeks

G 9.5 weeks

H 8.1 weeks

J 8 weeks

Holt Mathematics

Name _____ Date _____ Class _____

LESSON 3-8 Problem Solving
Interpret the Quotient

Write the correct answer.

1. Five friends split a pizza that costs $16.75. If they shared the bill equally, how much did they each pay?

2. There are 45 choir members going to the recital. Each van can carry 8 people. How many vans are needed?

3. Tara bought 150 beads. She needs 27 beads to make each necklace. How many necklaces can she make?

4. Cat food costs $2.85 for five cans. Ben only wants to buy one can. How much will it cost?

Circle the letter of the correct answer.

5. Tennis balls come in cans of 3. The coach needs 50 tennis balls for practice. How many cans should he order?

   A 16 cans
   B 17 cans
   C 18 cans
   D 20 cans

6. The rainfall for three months was 4.6 inches, 3.5 inches, and 4.2 inches. What was the average monthly rainfall during that time?

   F 41 inches
   G 12.3 inches
   H 4.3 inches
   J 4.1 inches

7. Tom has $15.86 to buy marbles that cost $1.25 each. He wants to know how many marbles he can buy. What should he do after he divides?

   A Drop the decimal part of the quotient when he divides.
   B Drop the decimal part of the dividend when he divides.
   C Round the quotient up to the next highest whole number to divide.
   D Use the entire quotient of his division as the answer.

8. Mei needs 135 hot dog rolls for the class picnic. The rolls come in packs of 10. She wants to know how many packs to buy. What should she do after she divides?

   F Drop the decimal part of the quotient when she divides.
   G Drop the decimal part of the dividend when she divides.
   H Round the quotient up to the next highest whole number.
   J Use the entire quotient of her division as the answer.

Holt Mathematics

Name _____ Date _____ Class _____

Problem Solving
Solving Decimal Equations

Write the correct answer.

1. Bee hummingbirds weigh only
   0.0056 ounces. They have to eat half
   their body weight every day to
   survive. How much food does a bee
   hummingbird have to eat each day?

   _____

2. The desert locust, a type of
   grasshopper, can jump 10 times
   the length of its body. The locust is
   1.956 inches long. How far can it
   jump in one leap?

   _____

3. In 1900, there were about
   1.49 million people living in
   California. In 2000, the population
   was 33.872 million. How much did
   the population grow between 1900
   and 2000?

   _____

4. Juanita has $567.89 in her checking
   account. After she deposited her
   paycheck and paid her rent of
   $450.00, she had $513.82 left in
   the account. How much was her
   paycheck?

   _____

Circle the letter of the correct answer.

5. The average body temperature for
   people is 98.6°F. The average body
   temperature for most dogs is 3.4°F
   higher than for people. The average
   body temperature for cats is 0.5°F
   lower than for dogs. What is the
   normal body temperature for dogs
   and cats?

   A dogs: 101.5°F; cats 102°F

   B dogs: 102°F; cats 101.5°F

   C dogs: 102.5°F; cats 103°F

   D dogs: 102.5°F; cats 102.5°F

6. Seattle, Washington, is famous for its
   rainy climate. Winter is the rainiest
   season there. From November
   through December the city gets an
   average of 5.85 inches of rain each
   month. Seattle usually gets 6 inches
   of rain in December. What is the
   city's average rainfall in November?

   F 6 inches

   G 5.925 inches

   H 5.8 inches

   J 5.7 inches

7. The equation to convert from Celsius
   to Kelvin degrees is K = 273.16 + C.
   If it is 303.66°K outside, what is the
   temperature in Celsius degrees?

   A 576.82°C

   B 30.5°C

   C 305°C

   D 257.68°C

8. The distance around a square mirror
   is 6.8 feet. Which of the following
   equations finds the length of each
   side of the mirror?

   F $6.8 - x = 4$

   G $x \div 4 = 6.8$

   H $4x = 6.8$

   J $6.8 + 4 = x$

Holt Mathematics

Problem Solving

LESSON 4-1 *Divisibility*

Use the table to answer the questions.

1. Which city's subway has a length that is a prime number of miles?

_____

2. Which subway could be evenly broken into sections of 2 miles each?

_____

3. Which subways could be evenly broken into sections of 5 miles each?

_____

_____

Subways Around the World

City, Country	Length (mi)
New York, U.S.	247
Mexico City, Mexico	111
Paris, France	125
Moscow, Russia	152
Seoul, South Korea	83
Tokyo, Japan	105

Circle the letter of the correct answer.

4. Which subway's length is divisible by 4 miles?

A New York, United States

B Paris, France

C Tokyo, Japan

D Moscow, Russia

5. Which subway's length is not a prime number, but is also not divisible by 2, 3, 4, 5, 6, or 9?

F Mexico City, Mexico

G New York, United States

H Seoul, South Korea

J Paris, France

6. The subway in Hong Kong, China, has a length that is a prime number of miles. Which of the following is its length?

A 260 miles

B 268 miles

C 269 miles

D 265 miles

7. The subway in St. Petersburg, Russia, has a length that is divisible by 3 miles. Which of the following is its length?

F 57 miles

G 56 miles

H 55 miles

J 58 miles

Holt Mathematics

Problem Solving
Factors and Prime Factorization

Write the correct answer.

1. The area of a rectangle is the product of its length and width. If a rectangular board has an area of 30 square feet, what are the possible measurements of its length and width?

2. The first-floor apartments in Jenna's building are numbered 100 to 110. How many apartments on that floor are a prime number? What are those apartment numbers?

3. A Russian mathematician named Christian Goldbach came up with a theory that every even number greater than 4 can be written as the sum of two odd primes. Test Goldbach's theory with the numbers 6 and 50.

4. Mr. Samuels has 24 students in his math class. He wants to divide the students into equal groups, and he wants the number of students in each group to be prime. What are his choices for group sizes? How many groups can he make?

Circle the letter of the correct answer.

5. Why is 2 the only even prime number?

   A It is the smallest prime number.

   B All other even numbers are divisible by 2.

   C It only has 1 and 2 as factors.

   D All odd numbers are prime.

6. What prime numbers are factors of both 60 and 105?

   F 2 and 3

   G 2 and 5

   H 3 and 5

   J 5 and 7

7. If a composite number has the first five prime numbers as factors, what is the smallest number it could be? Write that number's prime factorization.

   A 30

   B 210

   C 2,310

   D 30,030

8. Tim's younger brother, Bryant, just had a birthday. Bryant's age only has one factor, and is not a prime number. How old is Bryant?

   F 10 years old

   G 7 years old

   H 3 years old

   J 1 year old

Holt Mathematics

Problem Solving
LESSON 4-3 *Greatest Common Factor*

Write the correct answer.

1. Carolyn has 24 bottles of shampoo, 36 tubes of hand lotion, and 60 bars of lavender soap to make gift baskets. She wants to have the same number of each item in every basket. What is the greatest number of baskets she can make without having any of the items left over?

2. There are 40 girls and 32 boys who want to participate in the relay race. If each team must have the same number of girls and boys, what is the greatest number of teams that can race? How many boys and girls will be on each team?

3. Ming has 15 quarters, 30 dimes, and 48 nickels. He wants to group his money so that each group has the same number of each coin. What is the greatest number of groups he can make? How many of each coin will be in each group? How much money will each group be worth?

4. A gardener has 27 tulip bulbs, 45 tomato plants, 108 rose bushes, and 126 herb seedlings to plant in the city garden. He wants each row of the garden to have the same number of each kind of plant. What is the greatest number of rows that the gardener can make if he uses all the plants?

Circle the letter of the correct answer.

5. Kim packed 6 boxes with identical supplies. It was the greatest number she could pack and use all the supplies. Which of these is her supply list?

   A 24 pencils, 36 pens, 10 rulers

   B 12 rulers, 30 pencils, 45 pens

   C 42 pencils, 18 rulers, 72 pens

   D 60 pens, 54 pencils, 32 rulers

6. The sum of three numbers is 60. Their greatest common factor is 4. Which of the following lists shows those three numbers?

   F 4, 16, 36

   G 8, 20, 32

   H 14, 16, 30

   J 10, 18, 32

Holt Mathematics

Name _____ Date _____ Class _____

Problem Solving
Decimals and Fractions

Electricity is measured in amperes, or the rate electrical currents flow. A high ampere measurement means that a lot of electricity is being used. The table below shows the average amount of electricity some household appliances use per hour. Use the table to answer the questions.

1. How much electricity does an average 25-inch television use each hour? Write your answer as a decimal.

   _____

2. Which appliance uses an average of 2.5 amps per hour?

   _____

3. Which appliance uses the most electricity per hour? Write its ampere measurement as a decimal.

   _____

   _____

Electricity Use in the Home

Appliance	Amps per Hour
Blender	$2\frac{1}{2}$
Coffeemaker	$6\frac{2}{3}$
Computer and printer	$1\frac{5}{6}$
Microwave oven	$12\frac{1}{2}$
Popcorn popper	$2\frac{1}{12}$
25-inch television	$1\frac{1}{4}$
VCR	$\frac{1}{3}$

Circle the letter of the correct answer.

4. How much electricity do most computers and printers use in an hour?

   A 1.38 amperes

   B 1.8 amperes

   C $1.8\overline{3}$ amperes

   D 1.88 amperes

5. Which of the appliances has an hourly ampere measurement that is a repeating decimal?

   F blender

   G coffee maker

   H microwave oven

   J 25-inch television

6. In most years, 39.7 percent of the world's energy comes from burning oil. What is this percent written as a fraction?

   A $\frac{39}{7}$ percent

   B $39\frac{1}{7}$ percent

   C $3\frac{9}{7}$ percent

   D $39\frac{7}{10}$ percent

7. The United States produces about 13.2 percent of the world's hydroelectric power. What fraction of hydroelectric power does the United States produce?

   F $13\frac{1}{5}$ percent

   G $\frac{13}{2}$ percent

   H $1\frac{3}{2}$ percent

   J $13\frac{1}{2}$ percent

Holt Mathematics

Name _____ Date _____ Class _____

LESSON 4-5 Problem Solving
Equivalent Fractions

About 60 million Americans exercise 100 times or more each year. Their top activities and the fraction of those 60 million people who did them are shown on the circle graph. Use the graph to answer the questions.

Exercise in the U.S.

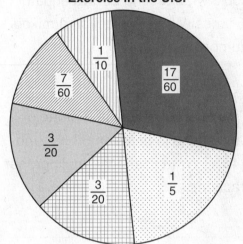

1. Which two activities did the same number of people use to keep in shape?

   _____

   _____

2. Which activity had the most participants? Write an equivalent fraction for that activity's participants.

   _____

   _____

3. Which activity had the fewest participants? Write two equivalent fractions for that activity's participants.

   _____

   _____

Legend:
- Fitness walking
- Free weights
- Stationary bike
- Running/Jogging
- Treadmill
- Resistance machines

Circle the letter of the correct answer.

4. Which activity did $\frac{3}{15}$ of the people use to exercise?

   A free weights

   B treadmill

   C fitness walking

   D stationary bike

5. Which activity did $\frac{35}{300}$ of the people use to stay healthy?

   F running/jogging

   G resistance machines

   H free weights

   J treadmill

6. An average-sized person can burn about $6\frac{1}{2}$ calories a minute while riding a bike. Which of the following is equivalent to that amount?

   A $1\frac{2}{2}$ C $6\frac{2}{4}$

   B $5\frac{6}{2}$ D $6\frac{2}{6}$

7. An average-sized person can burn about 11.25 calories a minute while jogging. Which of the following is not equivalent to that amount?

   F $11\frac{1}{4}$ H $11\frac{2}{8}$

   G $11\frac{1}{2}$ J $11\frac{3}{12}$

Holt Mathematics

Problem Solving

LESSON 4-6

Mixed Numbers and Improper Fractions

Write the correct answer.

1. If stretched end-to-end, the total length of the blood vessels inside your body could wrap around Earth's equator $\frac{5}{2}$ times! Write this fact as a mixed number.

2. In 2000, the average 12-year-old child in the United States earned an allowance of 9 dollars and $\frac{7}{25}$ cents a week. Write this amount as an improper fraction and a decimal.

3. The normal body temperature for a rattlesnake is between $53\frac{3}{5}°F$ and $64\frac{2}{5}°F$. Write this range as improper fractions.

4. A professional baseball can weigh no less than $\frac{45}{9}$ ounces and no more than $\frac{21}{4}$ ounces. Write this range as mixed numbers.

Circle the letter of the correct answer.

5. Betty needs a piece of lumber that is $\frac{14}{3}$ feet long. Which size should she look for at the hardware store?

   A $3\frac{1}{3}$ feet

   B $3\frac{1}{4}$ feet

   C $4\frac{2}{3}$ feet

   D $4\frac{1}{4}$ feet

6. What operations are used to change a mixed number to an improper fraction?

   F multiplication and addition

   G division and subtraction

   H division and addition

   J multiplication and subtraction

7. Adult bees only eat nectar, the substance in flowers used to make honey. A bee could fly 4 million miles on the energy it would get from eating $\frac{9}{2}$ liters of nectar. What is this amount of nectar written as a mixed number.

   A $9\frac{1}{2}$ liters   **C** $4\frac{1}{9}$ liters

   B $4\frac{1}{2}$ liters   **D** $2\frac{1}{2}$ liters

8. An astronaut who weighs 250 pounds on Earth would weigh $41\frac{1}{2}$ pounds on the moon. What is the astronaut's moon weight written as an improper fraction?

   F $\frac{41}{2}$ pounds   **H** $\frac{82}{2}$ pounds

   G $\frac{42}{2}$ pounds   **J** $\frac{83}{2}$ pounds

Holt Mathematics

Problem Solving

LESSON 4-7

Comparing and Ordering Fractions

The table shows what fraction of Earth's total land area each of the continents makes up. Use the table to answer the questions.

Earth's Land

Continent	Fraction of Earth's Land
Africa	$\frac{1}{5}$
Antarctica	$\frac{1}{10}$
Asia	$\frac{3}{10}$
Australia	$\frac{1}{20}$
Europe	$\frac{7}{100}$
North America	$\frac{4}{25}$
South America	$\frac{6}{50}$

1. Which continent makes up most of Earth's land?

   _____

2. Which continent makes up the least part of Earth's land?

   _____

3. Explain how you would compare the part of Earth's total land area that Australia and Europe make up.

   _____

   _____

Circle the letter of the correct answer.

4. Which of these continents covers the greatest part of Earth's total land area?

   A North America

   B South America

   C Europe

   D Australia

5. Which of these continents covers the least part of Earth's total land area?

   F Africa

   G Antarctica

   H Asia

   J Australia

6. Which of the following lists shows the continents written in order from the greatest part of Earth's total land they cover to the least part?

   A Asia, Africa, North America

   B Africa, Asia, North America

   C Asia, South America, North America

   D North America, Asia, South America

7. Which of the following lists shows the continents written in order from the least part of Earth's total land they cover to the greatest part?

   F Antarctica, Europe, South America

   G South America, Antarctica, Europe

   H Australia, Europe, Antarctica

   J Antarctica, Europe, Australia

Holt Mathematics

LESSON
4-8
Problem Solving
Adding and Subtracting with Like Denominators

Write the answers in simplest form.

1. About $\frac{3}{10}$ of Earth's surface is covered by land, and the rest is water. What fraction of Earth's surface is covered by water?

_____

2. A recipe for cookies calls for $\frac{3}{8}$ cup of chocolate chips. Tameeka wants to double the recipe. How much chocolate chips will she use?

_____

3. In Mr. Chesterfield's science class, $\frac{2}{9}$ of the boys like his class. Three times as many girls like his science class. How many girls like Mr. Chesterfield's science class?

_____

4. In the United States, $\frac{6}{50}$ of the population is left-handed men and $\frac{5}{50}$ of the population is left-handed women. What part of the population is left-handed?

_____

Circle the letter of the correct answer.

5. In the United States, about $\frac{1}{10}$ of the population is born with black hair, and $\frac{7}{10}$ of the population is born with brown hair. What fraction of the total population in the U.S. is born with brown or black hair?

A $\frac{1}{10}$ **C** $\frac{3}{5}$

B $\frac{1}{5}$ **D** $\frac{4}{5}$

6. In the United States, about $\frac{3}{20}$ of the population is born with blond hair, and $\frac{1}{20}$ of the population is born with red hair. What fraction of the total population in the U.S. is born with blond or red hair?

F $\frac{1}{10}$ **H** $\frac{3}{5}$

G $\frac{1}{5}$ **J** $\frac{4}{5}$

7. The average height for men in the United States is $5\frac{2}{3}$ feet tall. Bill is $\frac{1}{3}$ foot shorter than average. How tall is Bill?

A $5\frac{1}{3}$ feet **C** 6 feet

B $5\frac{3}{6}$ feet **D** $5\frac{1}{6}$ feet

8. The average height for women in the United States is $5\frac{1}{3}$ feet tall. Katie is $\frac{2}{3}$ foot taller than average. How tall is Katie?

F $5\frac{1}{3}$ feet **H** 6 feet

G $5\frac{3}{6}$ feet **J** $5\frac{1}{6}$ feet

Holt Mathematics

Name _____ Date _____ Class _____

Problem Solving
Estimating Fraction Sums and Differences

Use the table to answer the questions.

Portland, Oregon, Average Monthly Rainfall

Month	Jan	Feb	Mar	Apr	May	Jun	Jul	Aug	Sep	Oct	Nov	Dec
Rain (in.)	$5\frac{2}{5}$	$3\frac{9}{10}$	$3\frac{3}{5}$	$2\frac{2}{5}$	$2\frac{1}{10}$	$1\frac{1}{2}$	$\frac{3}{5}$	$1\frac{1}{10}$	$1\frac{4}{5}$	$2\frac{7}{10}$	$5\frac{3}{10}$	$6\frac{1}{10}$

1. About how much does it rain in Portland in January and February combined?

2. About how much more does it rain in Portland in October than in September?

_____

3. In most years, about how much rain does Portland receive from May through July?

4. What is the difference between Portland's average rainfall in March and May?

_____

Circle the letter of the correct answer.

5. What is the difference in rainfall between Portland's rainiest and driest months?

   A about $2\frac{1}{2}$ inches

   B about 5 inches

   C about $6\frac{1}{2}$ inches

   D about $7\frac{1}{2}$ inches

6. About how much rain does Portland receive in most years all together?

   F about $25\frac{1}{2}$ inches

   G about $30\frac{1}{2}$ inches

   H about $32\frac{1}{2}$ inches

   J about $36\frac{1}{2}$ inches

7. About how much rain does Portland receive during its three rainiest months all together?

   A about 17 inches

   B about 16 inches

   C about 18 inches

   D about 15 inches

8. In which month in Portland can you expect about $\frac{1}{2}$ inch less rainfall than in June?

   F May

   G July

   H September

   J August

Holt Mathematics

Problem Solving

LESSON 5-1

Least Common Multiple

Use the table to answer the questions.

1. You want to have an equal number of plastic cups and paper plates. What is the least number of packs of each you can buy?

   _____

   _____

2. You want to invite 48 people to a party. What is the least number of packs of invitations and napkins you should buy to have one for each person and none left over?

   _____

   _____

Party Supplies

Item	Number per Pack
Invitations	12
Balloons	30
Paper plates	10
Paper napkins	24
Plastic cups	15
Noise makers	5

Circle the letter of the correct answer.

3. You want to have an equal number of noisemakers and balloons at your party. What is the least number of packs of each you can buy?

   A 1 pack of balloons and 1 pack of noise makers

   B 1 pack of balloons and 2 packs of noise makers

   C 1 pack of balloons and 6 packs of noise makers

   D 6 packs of balloons and 1 pack of noise makers

5. The LCM for three items listed in the table is 60 packs. Which of the following are those three items?

   A balloons, plates, noise makers

   B noise makers, invitations, balloons

   C napkins, cups, plates

   D balloons, napkins, plates

4. You bought an equal number of packs of plates and cups so that each of your 20 guests would have 3 cups and 2 plates. How many packs of each item did you buy?

   F 1 pack of cups and 1 pack of plates

   G 3 packs of cups and 4 packs of plates

   H 4 packs of cups and 3 packs of plates

   J 4 packs of cups and 4 packs of plates

6. To have one of each item for 120 party guests, you buy 10 packs of one item and 24 packs of the other. What are those two items?

   F plates and invitations

   G balloons and cups

   H napkins and plates

   J invitations and noise makers

Holt Mathematics

Name _____ Date _____ Class _____

Use the circle graph to answer the questions. Write each answer in simplest form.

1. On which two continents do most people live? How much of the total population do they make up together?

   _____

   _____

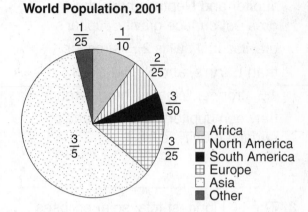

World Population, 2001

2. How much of the world's population live in either North America or South America?

   _____

3. How much more of the world's total population lives in Asia than in Africa?

   _____

Circle the letter of the correct answer.

4. How much of Earth's total population do people in Asia and Africa make up all together?

   A $\frac{3}{10}$ of the population

   B $\frac{2}{5}$ of the population

   C $\frac{7}{10}$ of the population

   D $\frac{7}{5}$ of the population

5. What is the difference between North America's part of the total population and Africa's part?

   F Africa has $\frac{1}{50}$ more.

   G Africa has $\frac{1}{50}$ less.

   H Africa has $\frac{9}{50}$ more.

   J Africa has $\frac{9}{50}$ less.

6. How much more of the population lives in Europe than in North America?

   A $\frac{1}{25}$ of the population

   B $\frac{1}{5}$ of the population

   C $\frac{1}{15}$ of the population

   D $\frac{1}{10}$ of the population

7. How much of the world's population lives in North America and Europe?

   F $\frac{1}{25}$ of the population

   G $\frac{1}{15}$ of the population

   H $\frac{1}{5}$ of the population

   J $\frac{1}{20}$ of the population

Holt Mathematics

Name _____ Date _____ Class _____

Problem Solving
Adding and Subtracting Mixed Numbers

Write the correct answer in simplest form.

1. Of the planets in our solar system, Jupiter and Neptune have the greatest surface gravity. Jupiter's gravitational pull is $2\frac{16}{25}$ stronger than Earth's, and Neptune's is $1\frac{1}{5}$ stronger. What is the difference between Jupiter's and Neptune's surface gravity levels?

2. Escape velocity is the speed a rocket must attain to overcome a planet's gravitational pull. Earth's escape velocity is $6\frac{9}{10}$ miles per second! The Moon's escape velocity is $5\frac{2}{5}$ miles per second slower. How fast does a rocket have to launch to escape the moon's gravity?

3. The two longest total solar eclipses occurred in 1991 and 1992. The first one lasted $6\frac{5}{6}$ minutes. The eclipse of 1992 lasted $5\frac{1}{3}$ minutes. How much longer was 1991's eclipse?

4. The two largest meteorites found in the U.S. landed in Canyon Diablo, Arizona, and Willamette, Oregon. The Arizona meteorite weighs $33\frac{1}{10}$ tons! Oregon's weighs $16\frac{1}{2}$ tons. How much do the two meteorites weigh in all?

Circle the letter of the correct answer.

5. Not including the Sun, Proxima Centauri is the closest star to Earth. It is $4\frac{11}{50}$ light years away! The next closest star is Alpha Centauri. It is $\frac{13}{100}$ light years farther than Proxima. How far is Alpha Centauri from Earth?

   A $4\frac{7}{20}$ light years

   B $4\frac{13}{100}$ light years

   C $4\frac{6}{25}$ light years

   D $4\frac{1}{50}$ light years

6. It takes about $5\frac{1}{3}$ minutes for light from the Sun to reach Earth. The Moon is closer to Earth, so its light reaches Earth faster—about $5\frac{19}{60}$ minutes faster than from the Sun. How long does light from the Moon take to reach Earth?

   F $\frac{3}{10}$ of a minute

   G $\frac{1}{60}$ of a minute

   H $\frac{1}{3}$ of a minute

   J $\frac{4}{15}$ of a minute

Holt Mathematics

Name _____ Date _____ Class _____

Write the correct answer in simplest form.

1. The average person in the United States eats $6\frac{13}{16}$ pounds of potato chips each year. The average person in Ireland eats $5\frac{15}{16}$ pounds. How much more potato chips do Americans eat a year than people in Ireland?

_____

2. The average person in the United States eats $270\frac{1}{16}$ pounds of meat each year. The average person in Australia eats $238\frac{1}{2}$ pounds. How much more meat do Americans eat a year than people in Australia?

_____

3. The average Americans eats $24\frac{1}{2}$ pounds of ice cream every year. The average person in Israel eats $15\frac{4}{5}$ pounds. How much more ice cream do Americans eat each year?

_____

4. People in Switzerland eat the most chocolate—26 pounds a year per person. Most Americans eat $12\frac{9}{16}$ pounds each year. How much more chocolate do the Swiss eat?

_____

5. The average person in the United States chews $1\frac{9}{16}$ pounds of gum each year. The average person in Japan chews $\frac{7}{8}$ pound. How much more gum do Americans chew?

_____

6. Norwegians eat the most frozen foods—$78\frac{1}{2}$ pounds per person each year. Most Americans eat $35\frac{15}{16}$ pounds. How much more frozen foods do people in Norway eat?

_____

Circle the letter of the correct answer.

7. Most people around the world eat $41\frac{7}{8}$ pounds of sugar each year. Most Americans eat $66\frac{3}{4}$ pounds. How much more sugar do Americans eat than the world's average?

   A $25\frac{7}{8}$ pounds more

   B $25\frac{1}{8}$ pounds more

   C $24\frac{7}{8}$ pounds more

   D $24\frac{1}{8}$ pounds more

8. The average person eats 208 pounds of vegetables and $125\frac{5}{8}$ pounds of fruit each year. How much more vegetables do most people eat than fruit?

   F $83\frac{5}{8}$ pounds more

   G $82\frac{3}{8}$ pounds more

   H $123\frac{5}{8}$ pounds more

   J $83\frac{3}{8}$ pounds more

Holt Mathematics

LESSON 5-5

Problem Solving
Solving Fraction Equations: Addition and Subtraction

Write the correct answer in simplest form.

1. It usually takes Brian $1\frac{1}{2}$ hours to get to work from the time he gets out of bed. His drive to the office takes $\frac{3}{4}$ hour. How much time does he spend getting ready for work?

2. Before she went to the hairdresser, Sheila's hair was $7\frac{1}{4}$ inches long. When she left the salon, it was $5\frac{1}{2}$ inches long. How much of her hair did Sheila get cut off?

3. One lap around the gym is $\frac{1}{3}$ mile long. Kim has already run 5 times around. If she wants to run 2 miles total, how much farther does she have to go?

4. Darius timed his speech at $5\frac{1}{6}$ minutes. His time limit for the speech is $4\frac{1}{2}$ minutes. How much does he need to cut from his speech?

Circle the letter of the correct answer.

5. Mei and Alex bought the same amount of food at the deli. Mei bought $1\frac{1}{4}$ pounds of turkey and $1\frac{1}{3}$ pounds of cheese. Alex bought $1\frac{1}{2}$ pounds of turkey. How much cheese did Alex buy?

   A $1\frac{1}{12}$ pounds   C $1\frac{1}{4}$ pounds

   B $1\frac{1}{6}$ pounds   D $4\frac{1}{12}$ pounds

6. When Lynn got her dog, Max, he weighed $10\frac{1}{2}$ pounds. During the next 6 months, he gained $8\frac{4}{5}$ pounds. At his one-year check-up he had gained another $4\frac{1}{3}$ pounds. How much did Max weigh when he was 1 year old?

   F $22\frac{19}{30}$ pounds   H $23\frac{29}{30}$ pounds

   G $23\frac{19}{30}$ pounds   J $23\frac{49}{50}$ pounds

7. Charlie picked up 2 planks of wood at the hardware store. One is $6\frac{1}{4}$ feet long and the other is $5\frac{5}{8}$ feet long. How much should he cut from the first plank to make them the same length?

   A $\frac{5}{8}$ foot   C $1\frac{3}{8}$ feet

   B $\frac{1}{2}$ foot   D $1\frac{5}{8}$ feet

8. Carmen used $3\frac{3}{4}$ cups of flour to make a cake. She had $\frac{1}{2}$ cup of flour left over. Which equation can you use to find how much flour she had before baking the cake?

   F $x + \frac{1}{2} = 3\frac{3}{4}$ H $3\frac{3}{4} - \frac{1}{2} = x$

   G $x - 3\frac{3}{4} = \frac{1}{2}$ J $3\frac{3}{4} - x = \frac{1}{2}$

Holt Mathematics

Name _____ Date _____ Class _____

Problem Solving
Multiplying Fractions by Whole Numbers

Write the answers in simplest form.

1. Did you know that some people have more bones than the rest of the population? About $\frac{1}{20}$ of all people have an extra rib bone. In a crowd of 60 people, about how many people are likely have an extra rib bone?

2. The Appalachian National Scenic Trail is the longest marked walking path in the United States. It extends through 14 states for about 2,000 miles. Last year, Carla hiked $\frac{1}{5}$ of the trail. How many miles of the trail did she hike?

3. Human fingernails can grow up to $\frac{1}{10}$ of a millimeter each day. How much can fingernails grow in one week?

4. Most people dream about $\frac{1}{4}$ of the time they sleep. How long will you probably dream tonight if you sleep for 8 hours?

Circle the letter of the correct answer.

5. Today, the United States flag has 50 stars—one for each state. The first official U.S. flag was approved in 1795. It had $\frac{3}{10}$ as many stars as today's flag. How many stars were on the first official U.S. flag?

   A 5 stars
   B 10 stars
   C 15 stars
   D 35 stars

6. The Statue of Liberty is about 305 feet tall from the ground to the tip of her torch. The statue's pedestal makes up about $\frac{1}{2}$ of its height. About how tall is the pedestal of the Statue of Liberty?

   F 610 feet
   G 152 1/2 feet
   H 150 1/2 feet
   J 102 1/2 feet

7. The Caldwells own a 60-acre farm. They planted $\frac{3}{5}$ of the land with corn. How many acres of corn did they plant?

   A 12 acres
   B 36 acres
   C 20 acres
   D 18 acres

8. Objects on Uranus weigh about $\frac{4}{5}$ of their weight on Earth. If a dog weighs 40 pounds on Earth, how much would it weigh on Uranus?

   F 32 pounds
   G 10 pounds
   H 8 pounds
   J 30 pounds

Holt Mathematics

LESSON **Problem Solving**
5-7 *Multiplying Fractions*

Use the circle graph to answer the questions. Write each answer in simplest form.

1. Of the students playing stringed instruments, $\frac{3}{4}$ play the violin. What fraction of the whole orchestra is violin players?

_____

2. Of the students playing woodwind instruments, $\frac{1}{2}$ play the clarinet. What fraction of the whole orchestra is clarinet players?

_____

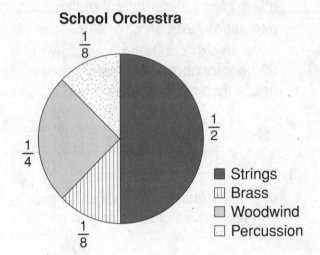

School Orchestra

$\frac{1}{8}$

$\frac{1}{2}$

$\frac{1}{4}$

$\frac{1}{8}$

■ Strings
▥ Brass
▨ Woodwind
□ Percussion

Circle the letter of the correct answer.

3. Two-thirds of the students who play a percussion instrument are boys. What fraction of the musicians in the orchestra is boys who play percussion? girls who play percussion?

   A $\frac{1}{24}$ of the orchestra

   B $\frac{1}{12}$ of the orchestra

   C $\frac{1}{4}$ of the orchestra

   D $\frac{2}{3}$ of the orchestra

4. The brass section is evenly divided into horns, trumpets, trombones, and tubas. What fraction of the whole orchestra do players of each of those brass instruments make up?

   F $\frac{1}{32}$ of the orchestra

   G $\frac{1}{8}$ of the orchestra

   H $\frac{1}{4}$ of the orchestra

   J $\frac{1}{2}$ of the orchestra

5. There are 40 students in the orchestra. How many students play either percussion or brass instruments?

   A 5 students

   B 10 students

   C 8 students

   D 16 students

6. If 2 more violinists join the orchestra, what fraction of all the musicians would play a stringed instrument?

   F $\frac{11}{21}$

   G $\frac{11}{20}$

   H $\frac{1}{20}$

   J $\frac{1}{26}$

Holt Mathematics

Name _____ Date _____ Class _____

LESSON **5-8**

Problem Solving
Multiplying Mixed Numbers

Use the recipe to answer the questions.

1. If you want to make $2\frac{1}{2}$ batches, how much flour would you need?

_____

2. If you want to make only $1\frac{1}{2}$ batches, how much chocolate chips would you need?

_____

3. You want to bake $3\frac{1}{4}$ batches. How much vanilla do you need in all?

_____

CHOCOLATE CHIP COOKIES
Servings: 1 batch
$1\frac{2}{3}$ cups flour
$\frac{3}{4}$ teaspoon baking soda
$\frac{1}{2}$ cup white sugar
$2\frac{1}{3}$ cups semisweet chocolate chips
$\frac{1}{2}$ cup brown sugar
$\frac{3}{4}$ cup butter
1 egg
$1\frac{1}{4}$ teaspoons vanilla

Choose the letter for the best answer.

4. If you make $1\frac{1}{4}$ batches, how much baking soda would you need?

   A $\frac{3}{16}$ teaspoon   C $\frac{3}{5}$ teaspoon

   B $\frac{5}{16}$ teaspoon   D $\frac{15}{16}$ teaspoon

5. How many cups of white sugar do you need to make $3\frac{1}{2}$ batches of cookies?

   F $3\frac{1}{2}$ cups   H $1\frac{1}{2}$ cups

   G $1\frac{3}{4}$ cups   J $1\frac{1}{4}$ cups

6. Dan used $2\frac{1}{4}$ cups of butter to make chocolate chip cookies using the above recipe. How many batches of cookies did he make?

   A 3 batches   C 5 batches

   B 4 batches   D 6 batches

7. One bag of chocolate chips holds 2 cups. If you buy five bags, how many cups of chips will you have left over after baking $2\frac{1}{2}$ batches of cookies?

   F $4\frac{1}{6}$ cups   H $2\frac{1}{3}$ cups

   G $5\frac{5}{6}$ cups   J $\frac{1}{3}$ cup

Holt Mathematics

Name _____ Date _____ Class _____

Problem Solving
Dividing Fractions and Mixed Numbers

Write the correct answer in simplest form.

1. Horses are measured in units called *hands*. One inch equals $\frac{1}{4}$ hand. The average Clydesdale horse is $17\frac{1}{5}$ hands high. What is the horse's height in inches? in feet?

   _____

2. Cloth manufacturers use a unit of measurement called a *finger*. One finger is equal to $4\frac{1}{2}$ inches. If 25 inches are cut off a bolt of cloth, how many fingers of cloth were cut?

   _____

3. People in England measure weights in units called *stones*. One pound equals $\frac{1}{14}$ of a stone. If a cat weighs $\frac{3}{4}$ stone, how many pounds does it weigh?

   _____

4. The hiking trail is $\frac{9}{10}$ mile long. There are 6 markers evenly posted along the trail to direct hikers. How far apart are the markers placed?

   _____

Choose the letter for the best answer.

5. A cake recipe calls for $1\frac{1}{2}$ cups of butter. One tablespoon equals $\frac{1}{16}$ cup. How many tablespoons of butter do you need to make the cake?

   A 24 tablespoons

   B 8 tablespoons

   C $\frac{3}{32}$ tablespoon

   D 9 tablespoons

6. Printed letters are measured in units called *points*. One point equals $\frac{1}{72}$ inch. If you want the title of a paper you are typing on a computer to be $\frac{1}{2}$ inch tall, what type point size should you use?

   F 144 point

   G 36 point

   H $\frac{1}{36}$ point

   J $\frac{1}{144}$ point

7. Phyllis bought 14 yards of material to make chair cushions. She cut the material into pieces $1\frac{3}{4}$ yards long to make each cushion. How many cushions did Phyllis make?

   A 4 cushions   **C** 8 cushions

   B 6 cushions   **D** $24\frac{1}{2}$ cushions

8. Dry goods are sold in units called *pecks* and *bushels*. One peck equals $\frac{1}{4}$ bushel. If Peter picks $5\frac{1}{2}$ bushels of peppers, how many pecks of peppers did Peter pick?

   F $1\frac{3}{8}$ pecks   **H** 20 pecks

   G 11 pecks   **J** 22 pecks

Holt Mathematics

LESSON 5-10 Problem Solving
Solving Fraction Equations: Multiplication and Division

Solve.

1. The number of T-shirts is multiplied by $\frac{1}{2}$ and the product is 18. Write and solve an equation for the number of T-shirts, where t represents the number of T-shirts.

2. The number of students is divided by 18 and the quotient is $\frac{1}{6}$. Write and solve an equation for the number of students, where s represents the number of students.

3. The number of players is multiplied by $2\frac{1}{2}$ and the product is 25. Write and solve an equation for the number of players, where p represents the number of players.

4. The number of chairs is divided by $\frac{1}{4}$ and the quotient is 12. Write and solve an equation for the number of chairs, where c represents the number of chairs.

Circle the letter of the correct answer.

5. Paco bought 10 feet of rope. He cut it into several $\frac{5}{6}$-foot pieces. Which equation can you use to find how many pieces of rope Paco cut?

   A $\frac{5}{6} \div 10 = x$

   B $\frac{5}{6} \div x = 10$

   C $10 \div x = \frac{5}{6}$

   D $10x = \frac{5}{6}$

6. Each square on the graph paper has an area of $\frac{4}{9}$ square inch. What is the length and width of each square?

   F $\frac{1}{9}$ inch

   G $\frac{2}{3}$ inch

   H $\frac{2}{9}$ inch

   J $\frac{1}{3}$ inch

7. Which operation should you use to solve the equation $6x = \frac{3}{8}$?

   A addition

   B subtraction

   C multiplication

   D division

8. A fraction divided by $\frac{2}{3}$ is equal to $1\frac{1}{4}$. What is that fraction?

   F $\frac{1}{3}$

   G $\frac{5}{6}$

   H $\frac{1}{4}$

   J $\frac{1}{2}$

Holt Mathematics

Problem Solving

Make a Table

Complete each activity and answer each question.

1. In January, the normal temperature in Atlanta, Georgia, is 41°F. In February, the normal temperature in Atlanta is 45°F. In March, the normal temperature in Atlanta is 54°F, and in April, it is 62°F. Atlanta's normal temperature in May is 69°F. Use this data to complete the table at right.

2. Use your table from Exercise 1 to find a pattern in the data and draw a conclusion about the temperature in June.

_____

_____

_____

3. In what other ways could you organize the data in a table?

_____

_____

Circle the letter of the correct answer.

4. In which month given does Atlanta have the highest temperature?

   A February

   B March

   C April

   D May

5. In which month given does Atlanta have the lowest temperature?

   F January

   G February

   H March

   J April

6. Which of these statements about Atlanta's temperature data from January to May is true?

   A It is always higher than 40°F.

   B It is always lower than 60°F.

   C It is hotter in March than in April.

   D It is cooler in February than in January.

7. Between which two months in Atlanta does the normal temperature change the most?

   F January and February

   G February and March

   H March and April

   J April and May

Holt Mathematics

LESSON 6-2 Problem Solving
Mean, Median, Mode, and Range

Write the correct answer.

1. Use the table at right to find the mean, median, mode, and range of the data set.

   _____

   _____

   _____

2. When you use the data for only 2 of the teams in the table, the mean, median, and mode for the data are the same. Which teams are they?

   _____

   _____

World Series Winners

Team	Number of Wins
Baltimore Orioles	3
Boston Red Sox	5
Detroit Tigers	4
Minnesota Twins	3
Pittsburgh Pirates	5

Circle the letter of the correct answer.

3. The states that border the Gulf of Mexico are Alabama, Florida, Louisiana, Mississippi, and Texas. What is the mean for the number of letters in those states' names?

   A 7 letters

   B 7.8 letters

   C 8 letters

   D 8.7 letters

4. There are 5 whole numbers in a data set. The mean of the data is 10. The median and mode are both 9. The least number in the data set is 7, and the greatest is 14. What are the numbers in the data set?

   F 7, 7, 9, 11, and 14

   G 7, 7, 9, 9, and 14

   H 7, 9, 9, 11, and 14

   J 7, 9, 9, 14, and 14

5. If the mean of two numbers is 2.5, what is true about the data?

   A Both numbers are greater than 5.

   B One of the numbers is less than 2.

   C One of the numbers is 2.5.

   D The sum of the data is not divisible by 2.

6. Tom wants to find the average height of the students in his class. Which measurement should he find?

   F the range

   G the mean

   H the median

   J the mode

Holt Mathematics

Problem Solving

Additional Data and Outliers

Use the table to answer the questions.

1. Find the mean, median, and mode of the earnings data.

_____

_____

2. *Titanic* earned more money in the United States than any other film—a total of $600 million! Add this figure to the data and find the mean, median, and mode. Round your answer for the mean to the nearest whole million.

_____

_____

Successful Films in the U.S.

Film	U.S. Earnings for first release (million $)
E.T. the Extra-Terrestrial	400
Forrest Gump	330
Independence Day	305
Jurassic Park	357
The Lion King	313

Circle the letter of the correct answer.

3. In Canada, people watch TV an average of 74 minutes each day. In Germany, people watch an average of 68 minutes a day. In France it is 67 minutes a day, in Spain it is 91 minutes a day, and in Ireland it is 74 minutes a day. Find the mean, median, and mode of the data.

   A mean: 74 min.; median: 74 min.; mode: 74 min.

   B mean: 74 min.; median: 74.8 min.; mode: 74 min.

   C mean: 74.8 min.; median: 74 min.; mode: 24 min.

   D mean: 74.8 min.; median: 74 min.; mode: 74 min.

5. In Exercise 2, which data measurement changed the least with the addition of *Titanic's* earnings?

   A the range   C the median

   B the mean   D the upper extreme

4. People in the United States watch more television than in any other country. Americans watch an average of 118 minutes a day! Add this number to the data and find the mean, median, and mode.

   F mean: 82 min.; median: 74 min.; mode: 74 min.

   G mean: 82 min.; median: 74 min.; mode: 118 min.

   H mean: 82 min.; median: 91 min.; mode: 74 min.

   J mean: 74.8 min.; median: 82 min.; mode: 74 min.

6. In Exercise 4, which measurements best describe the data?

   F mean and median

   G range and mean

   H median and mode

   J range and mode

Holt Mathematics

Name _____ Date _____ Class _____

LESSON 6-4

Problem Solving
Bar Graphs

Use the bar graph for Exercises 1–4.

1. What is the range of the goals the hockey players scored per season?

   _____

2. What is the mode of the goals scored?

   _____

3. What is the mean number of goals the players scored?

   _____

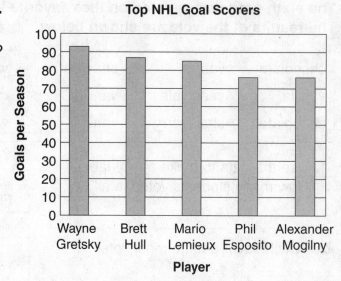

Top NHL Goal Scorers

Use the bar graph for Exercises 5–8.

4. Which team won the most games

   that season? _____

5. Which team lost the most games

   that season? _____

6. What was the mean number of

   games won? _____

7. What was the mean number of

   games lost? _____

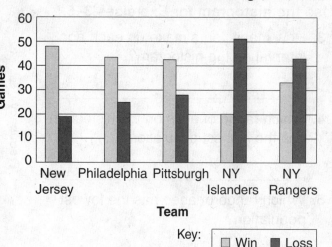

NHL Eastern Conference Final Standings, 2000–2001

Key: ☐ Win ■ Loss

Circle the letter of the correct answer.

8. Which hockey team had the greatest difference between the number of games won and lost?

   A New Jersey

   B New York Islanders

   C Philadelphia

   D Pittsburgh

9. How do you know the mode of a data set by looking at a bar graph?

   F The mode has two or more bars on the graph with the same height.

   G The mode has the tallest bar.

   H The mode has the lowest bar.

   J The bar for the mode is in the middle of the graph.

Holt Mathematics

Name _____ Date _____ Class _____

Problem Solving

LESSON 6-5

Line Plots, Frequency Tables, and Histograms

The sixth grade class voted on their favorite ice cream flavors. The results of the vote are shown below.

chocolate	vanilla	strawberry	vanilla	vanilla
vanilla	chocolate	vanilla	chocolate	strawberry
chocolate	strawberry	vanilla	vanilla	chocolate

1. Use the data to make a tally table. How many students voted in all?

_____

2. Which flavor got the most votes?

_____

Ice Cream Flavor Votes

Flavor	Number of Votes

Use the histogram for Exercises 3–5.

3. How many years make up each age interval on the histogram?

_____

4. Which range of ages on the histogram has the highest population?

_____

5. Which range of ages has the lowest population?

_____

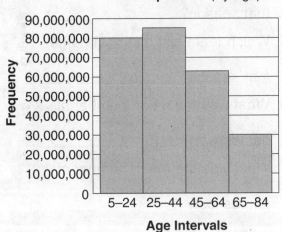

U.S. Population (By Age)

Circle the letter of the correct answer.

6. Which of the following cannot be used to make a frequency table with intervals?

A histogram

B tally table

C line plot

D double-bar graph

7. Which question can be answered by using the histogram above?

A How many people in the United States are younger than 5 years?

B What is the mean age of all people in the United States?

C How many people in the United States are older than 84 years old?

D How many people in the United States are age 25 to 64?

Holt Mathematics

Name _____ Date _____ Class _____

Use the coordinate grid to answer each question.

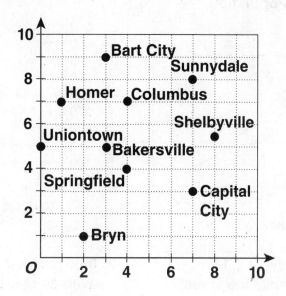

1. What city is located at point (4, 4) on the map?

_____

2. Which city is located at point $(8, 5\frac{1}{2})$ on the map?

_____

3. Which city's location is given by an ordered pair that includes a 0?

_____

4. What ordered pair describes the location of Capital City?

_____

5. If you started at (0, 0) and moved 1 unit north and 2 units east, which city would you reach?

6. Which two cities on the map are both located 4 units to the right of (0, 0)?

Circle the letter of the correct answer.

7. If you started in Bart City and moved 2 units south and 2 units west, which city would you reach?

   A Columbus

   B Sunnydale

   C Homer

   D Bakersville

8. Starting at (0, 0), which of the following directions would lead you to Capital City?

   F Go 7 units east and 3 units north.

   G Go 5 units north and 3 units east.

   H Go 3 unit east and 7 units north.

   J Go 8 units east and 6 units north.

Holt Mathematics

Problem Solving
LESSON 6-7 *Line Graphs*

Use the line graphs to answer each question.

U.S. Farm Population

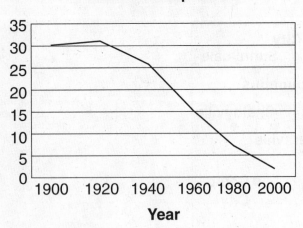

Size of U.S. Farms

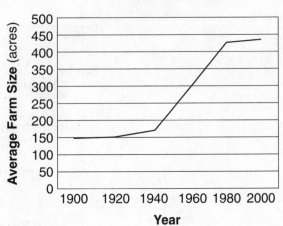

1. In which year was the U.S. farm population the highest? the lowest?

   _____

2. In which year was the size of the average U.S. farm the largest? the smallest?

   _____

3. In general, how has the U.S. farm population changed in the last 100 years?

   _____

4. In general, how has the size of the average U.S. farm changed in the last 100 years?

   _____

Circle the letter of the correct answer.

5. How many people lived on farms in the United States in 1940?

   A 31 million

   B 30 million

   C 26 million

   D 15 million

6. How many acres did the average farm in the United States cover in 1980?

   F 150 acres

   G 300 acres

   H 400 acres

   J 426 acres

7. Between which two years did the U.S. farm population increase?

   A 1900 and 1920

   B 1920 and 1940

   C 1940 and 1960

   D 1960 and 1980

8. Between which two years did the average size of farms in the United States change the least?

   F 1900 and 1920

   G 1920 and 1940

   H 1960 and 1980

   J 1980 and 2000

Holt Mathematics

Name _____ Date _____ Class _____

Problem Solving
Misleading Graphs

Use the graphs to answer each question.

Graph A

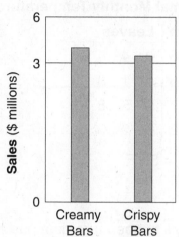

Graph B

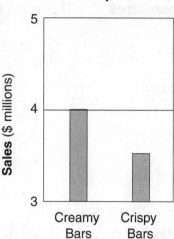

Graph C

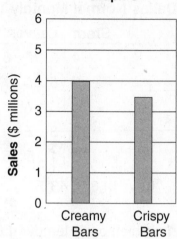

1. Why is Graph A misleading?

_____

_____

_____

2. Why is Graph B misleading?

_____

_____

_____

3. What might people believe from reading Graph A?

_____

_____

4. What might people believe from reading Graph B?

_____

_____

Circle the letter of the correct answer.

5. Which of the following information is different on all three graphs above?

  A the vertical scale

  B the Crispy Bars sales data

  C the Creamy Bars sales data

  D the horizontal scale

6. Which of the following is a way that graphs can be misleading?

  F breaks in scales

  G uneven scales

  H missing parts of scales

  J all of the above

7. Which graph do you think was made by the company that sells Crispy Bars?

  A Graph A   **C** Graph C

  B Graph B   **D** all of the graphs

8. If you were writing a newspaper article about candy bar sales, which graph would be best to use?

  F Graph A   **H** Graph C

  G Graph B   **J** all of the above

Holt Mathematics

Name _____ Date _____ Class _____

Problem Solving
Stem-and-Leaf Plots

Use the Texas stem-and-leaf plots to answer each question.

Dallas Normal Monthly Temperatures

Stem	Leaves
4	3 7 8
5	6 7
6	6 7
7	3 7
8	1 5 5

Key: 4 | 3 = 43°F

Houston Normal Monthly Temperatures

Stem	Leaves
5	0 4 4
6	1 1 8
7	0 5 8
8	0 2 3

Key: 5 | 0 = 50°F

1. Which city's temperature data has a mode of 85°F?

_____

2. Which city's temperature data has a range of 33°F?

_____

3. Which city has the lowest data value? What is that value?

_____

4. Which city has the highest data value? What is that value?

_____

Circle the letter of the correct answer.

5. Which city's temperature data has a mean of 68°F?

   A Dallas

   B Houston

   C both Dallas and Houston

   D neither Dallas nor Houston

6. Which city's temperature data has a median of 69°F?

   F Dallas

   G Houston

   H both Dallas and Houston

   J neither Dallas nor Houston

7. What do the data values 54°F and 61°F represent for the plots above?

   A the ranges of normal temperatures in Dallas and Houston

   B the mode of normal temperatures for Houston

   C the mean and median normal temperatures for Dallas

   D the lowest normal temperatures for Dallas and Houston

8. Which of the following would be the best way to display the Dallas and Houston temperature data?

   F on a line graph

   G in a tally table

   H on a bar graph

   J on a coordinate plane

Holt Mathematics

Problem Solving
LESSON 6-10 *Choosing an Appropriate Display*

1. Write *line plot, stem-and-leaf plot, line graph,* or *bar graph* to describe the most appropriate way to show the height of a sunflower plant every week for one month.

_____

2. Write *line plot, stem-and-leaf plot, line graph,* or *bar graph* to describe the most appropriate way to show the number of votes received by each candidate running for class president

_____

3. Write *line plot, stem-and-leaf plot, line graph,* or *bar graph* to describe the most appropriate way to show the test scores each student received on a math quiz.

_____

4. Write *line plot, stem-and-leaf plot, line graph,* or *bar graph* to describe the most appropriate way to show the average time spent sleeping per day by 30 sixth-grade students.

_____

Circle the letter of the correct answer.

5. People leaving a restaurant were asked how much they spent for lunch. Here are the results of the survey to the nearest dollar: $8, $7, $9, $7, $10, $5, $8, $8, $12, $8. Which type of graph would be most appropriate to show the data?

   A bar graph

   B line graph

   C line plot

   D stem-and-leaf plot

6. People leaving a movie theater were asked their age. Here are the results of the survey to the nearest year: 12, 11, 13, 15, 22, 31, 40, 12, 17, 20, 33, 16, 12, 24, 19. Which type of graph would be most appropriate to show the data?

   F bar graph

   G line graph

   H line plot

   J stem-and-leaf plot

7. What is the median amount of money spent on lunch in Exercise 5?

   A $7

   B $8

   C $9

   D $12

8. What is the median age of the movie-goers in Exercise 6?

   F 15

   G 16

   H 17

   J 19

Holt Mathematics

LESSON
7-1
Problem Solving
Ratios and Rates

Use the table to answer each question.

Atomic Particles of Elements

Element	Protons	Neutrons	Electrons
Gold	79	118	79
Iron	26	30	26
Neon	10	10	10
Platinum	78	117	78
Silver	47	61	47
Tin	50	69	50

1. What is the ratio of gold protons to silver protons?

_____

2. What is the ratio of gold neutrons to platinum protons?

_____

3. What are two equivalent ratios of the ratio of neon protons to tin protons?

_____

4. What are two equivalent ratios of the ratio of iron protons to iron neutrons?

_____

Circle the letter of the correct answer.

5. A ratio of one element's neutrons to another element's electrons is equivalent to 3 to 5. What are those two elements?

   A iron neutrons to tin electrons

   B gold neutrons to tin electrons

   C tin neutrons to gold electrons

   D neon neutrons to iron electrons

6. The ratio of two elements' protons is equivalent to 3 to 1. What are those two elements?

   F gold to tin

   G neon to tin

   H platinum to iron

   J silver to gold

7. Which element in the table has a ratio of 1 to 1, no matter what parts you are comparing in the ratio?

   A iron   C tin

   B neon   D silver

8. If the ratio for any element is 1:1, which two parts is the ratio comparing?

   F protons to neutrons

   G electrons to neutrons

   H protons to electrons

   J neutrons to electrons

Holt Mathematics

Name _____ Date _____ Class _____

Problem Solving

Using Tables to Explore Equivalent Ratios and Rates

Use the table to answer the questions.

School Outing Student-to-Parent Ratios

Number of Students	8	16	24	32	40	48	56	64	72
Number of Parents	2	4	6	8	10	12	14	16	18

1. Each time some students go on a school outing, their teachers invite students' parents to accompany them. Predict how many parents will accompany 88 students.

_____

2. Next week 112 students will go to the Science Museum. Their teachers invited some of the students' parents to go with them. How many parents do you predict will go with the students to the Science Museum?

_____

Circle the letter of the correct answer.

3. Tanya's class of 28 students will be going to the Nature Center. How many parents do you predict Tanya's teacher will invite to accompany them?

A 5 parents

B 7 parents

C 9 parents

D 11 parents

4. Some students will be going on an outing to the local police station. Their teachers invited 13 parents to accompany them. How many students do you predict will be going on the outing?

F 49 students

G 50 students

H 51 students

J 52 students

5. In June, all of the students in the school will be going on their annual picnic. If there are 416 students in the school, what do you predict the number of parents accompanying them on the picnic will be?

A 52 parents

B 78 parents

C 104 parents

D 156 parents

6. On Tuesday, all of the sixth-grade students will be going to the Space Museum. Their teachers invited 21 parents to accompany them. How many sixth graders do you predict will be going to the Space Museum?

F 80 sixth graders

G 82 sixth graders

H 84 sixth graders

J 86 sixth graders

Holt Mathematics

Problem Solving
Proportions

Write the correct answer.

1. For most people, the ratio of the length of their head to their total height is 1:7. Use proportions to test your measurements and see if they match this ratio.

_____

_____

2. The ratio of an object's weight on Earth to its weight on the Moon is 6:1. The first person to walk on the Moon was Neil Armstrong. He weighed 165 pounds on Earth. How much did he weigh on the Moon?

_____

_____

3. It has been found that the distance from a person's eye to the end of the fingers of his outstretched hand is proportional to the distance between his eyes at a 10:1 ratio. If the distance between your eyes is 2.3 inches, what should the distance from your eye to your outstretched fingers be?

_____

_____

4. Chemists write the formula of ordinary sugar as $C_{12}H_{22}O_{11}$, which means that the ratios of 1 molecule of sugar are always 12 carbon atoms to 22 hydrogen atoms to 11 oxygen atoms. If there are 4 sugar molecules, how many atoms of each element will there be?

_____

_____

Circle the letter of the correct answer.

5. A healthy diet follows the ratio for meat to vegetables of 2.5 servings to 4 servings. If you eat 7 servings of meat a week, how many servings of vegetables should you eat?

A 28 servings   **C** 14 servings

B 17.5 servings   **D** 11.2 servings

6. A 150-pound person will burn 100 calories while sitting still for 1 hour. Following this ratio, how many calories will a 100-pound person burn while sitting still for 1 hour?

F $666\frac{2}{3}$ calories   **H** $6\frac{2}{3}$ calories

G $66\frac{2}{3}$ calories   **J** 6 calories

7. Recently, 1 U.S. dollar was worth 1.58 in euros. If you exchanged $25 at that rate, how many euros would you get?

A 39.50 euros

B 15.82 euros

C 26.58 euros

D 23.42 euros

8. Recently, 1 U.S. dollar was worth 0.69 English pound. If you exchanged 500 English pounds, how many dollars would you get?

F 345 U.S. dollars

G 725 U.S. dollars

H 500.69 U.S dollars

J 499.31 U.S. dollars

Holt Mathematics

Name _____ Date _____ Class _____

LESSON 7-4

Problem Solving
Similar Figures

Write the correct answer.

1. The map at right shows the dimensions of the Bermuda Triangle, a region of the Atlantic Ocean where many ships and airplanes have disappeared. If a theme park makes a swimming pool in a similar figure, and the longest side of the pool is 0.5 mile long, about how long would the other sides of the pool have to be?

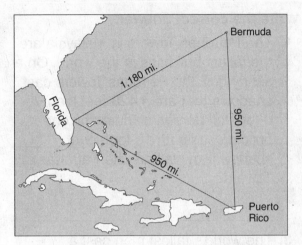

_____

2. Completed in 1883, *The Battle of Gettysburg* is 410 feet long and 70 feet tall. A museum shop sells a print of the painting that is similar to the original. The print is 2.05 feet long. How tall is the print?

3. *Panorama of the Mississippi* was 12 feet tall and 5,000 feet long! If you wanted to make a copy similar to the original that was 2 feet tall, how many feet long would the copy have to be?

_____

Circle the letter of the correct answer.

4. Two tables shaped like triangles are similar. The measure of one of the larger table's angles is 38°, and another angle is half that size. What are the measures of all the angles in the smaller table?

   A 19°, 9.5°, and 61.5°

   B 38°, 19°, and 123°

   C 38°, 38°, and 104°

   D 76°, 38°, and 246°

6. Which of the following is not always true if two figures are similar?

   A They have the same shape.

   B They have the same size.

   C Their corresponding sides have proportional lengths.

   D Corresponding angles are congruent.

5. Two rectangular gardens are similar. The area of the larger garden is 8.28 m², and its length is 6.9 m. The smaller garden is 0.6 m wide. What is the smaller garden's length and area?

   F length = 6.9 m; area = 2.07 m²

   G length = 3.45 m; area = 4.14 m²

   H length = 3.45 m; area = 1.97 m²

   J length = 3.45 m; area = 2.07 m²

7. Which of the following figures are always similar?

   F two rectangles

   G two triangles

   H two squares

   J two pentagons

Holt Mathematics

Problem Solving

LESSON 7-5

Indirect Measurement

Write the correct answer.

1. The Petronas Towers in Malaysia are the tallest buildings in the world. On a sunny day, the Petronas Towers cast shadows that are 4,428 feet long. A 6-foot-tall person standing by one building casts an 18-foot-long shadow. How tall are the Petronas Towers?

_____

2. The Sears Tower in Chicago is the tallest building in the United States. On a sunny day, the Sears Tower casts a shadow that is 2,908 feet long. A 5-foot-tall person standing by the building casts a 10-foot-long shadow. How tall is the Sears Tower?

_____

3. The world's tallest man cast a shadow that was 535 inches long. At the same time, a woman who was 5 feet 4 inches tall cast a shadow that was 320 inches long. How tall was the world's tallest man in feet and inches?

_____

4. Hoover Dam on the Colorado River casts a shadow that is 2,904 feet long. At the same time, an 18-foot-tall flagpole next to the dam casts a shadow that is 72 feet long. How tall is Hoover Dam?

_____

Circle the letter of the correct answer.

5. An NFL goalpost casts a shadow that is 170 feet long. At the same time, a yardstick casts a shadow that is 51 feet long. How tall is an NFL goalpost?

   A 100 feet

   B 56 2/3 feet

   C 10 feet

   D 1 foot

6. A gorilla casts a shadow that is 600 centimeters long. A 92-centimeter-tall chimpanzee casts a shadow that is 276 centimeters long. What is the height of the gorilla in meters?

   F 0.2 meter

   G 2 meters

   H 20 meters

   J 200 meters

7. A 6-foot-tall man casts a shadow that is 30 feet long. If a boy standing next to the man casts a shadow that is 12 feet long, how tall is the boy?

   A 2.2 feet   **C** 2.4 feet

   B 5 feet   **D** 2 feet

8. An ostrich is 108 inches tall. If its shadow is 162 inches, and an emu standing next to it casts a 90-inch shadow, how tall is the emu?

   F 162 inches   **H** 60 inches

   G 90 inches   **J** 194.4 inches

Holt Mathematics

Name _____ Date _____ Class _____

Problem Solving

Scale Drawings and Maps

Write the correct answer.

1. About how many kilometers long is the northern border of California along Oregon?

_____

2. What is the distance in kilometers from Los Angeles to San Francisco?

_____

3. How many kilometers would you have to drive to get from San Diego to Sacramento?

_____

4. At its longest point, about how many kilometers long is Death Valley National Park?

_____

5. Approximately what is the distance, in kilometers, between Redwood National Park and Yosemite National Park?

_____

Circle the letter of the correct answer.

6. Which of the following two cities in California are about 200 kilometers apart?

   A San Diego and Los Angeles

   B Monterey and Los Angeles

   C San Francisco and Fresno

   D Palm Springs and Bakersfield

7. Joshua Tree National Park is about 200 kilometers from Sequoia National Park. How many centimeters should separate those parks on this map?

   F 110 cm   H 1 cm

   G 11 cm   J 0.11 cm

Holt Mathematics

Name _____ Date _____ Class _____

Problem Solving
Percents

Use the circle graph to answer each question. Write fractions in simplest form.

1. What fraction of the total 2000 music sales in the United States were rock recordings?

_____

2. On this grid, model the percent of total United States music sales that were rap recordings. Then write that percent as a decimal.

_____

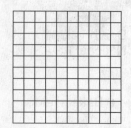

U.S. Recorded Music Sales, 2000

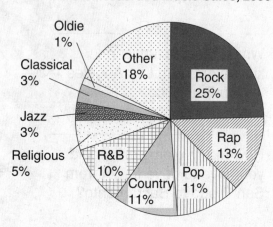

Circle the letter of the correct answer.

3. What kind of music made up $\frac{1}{20}$ of the total U.S. music recording sales?

   A Oldie   **C** Jazz

   B Classical   **D** Religious

4. What fraction of the United States music sales were country recordings?

   F $\frac{110}{100}$ **H** $\frac{1}{10}$

   G $\frac{11}{100}$ **J** $\frac{1}{100}$

5. What fraction of all United States recording sales did jazz and classical music make up together?

   A $\frac{6}{10}$ **C** $\frac{1}{5}$

   B $\frac{3}{50}$ **D** $\frac{11}{100}$

6. What kind of music made up $\frac{1}{10}$ of the total music recording sales in the United States in 2000?

   F Pop   **H** R&B

   G Jazz   **J** Oldies

Holt Mathematics

Name _____ Date _____ Class _____

LESSON 7-8 Problem Solving
Percents, Decimals, and Fractions

Write the correct answer.

1. Deserts cover about $\frac{1}{7}$ of all the land on Earth. About what percent of Earth's land is made up of deserts?

_____

2. The Sahara is the largest desert in the world. It covers about 3% of the total area of Africa. What decimal expresses this percent?

_____

3. Cactus plants survive in deserts by storing water in their thick stems. In fact, water makes up $\frac{3}{4}$ of the saguaro cactus's total weight. What percent of its weight is water?

_____

4. Daytime temperatures in the Sahara can reach 130°F! At night, however, the temperature can drop by 62%. What decimal expresses this percent?

_____

Circle the letter of the correct answer.

5. The desert nation of Saudi Arabia is the world's largest oil producer. About $\frac{1}{4}$ of all the oil imported to the United States is shipped from Saudi Arabia. What percent of our nation's oil is that?
   A 20%
   B 22%
   C 25%
   D 40%

6. About $\frac{2}{5}$ of all the food produced on Earth is grown on irrigated cropland. What percent of the world's food production relies on irrigation? What is the percent written as a decimal?
   F 40%; 40.0
   G 40%; 4.0
   H 40%; 0.4
   J 40%; 0.04

7. About $\frac{3}{25}$ of all the freshwater in the United States is used for drinking, washing, and other domestic purposes. What percent of our fresh water resources is that?
   A 3%
   B 25%
   C 12%
   D $\frac{1}{5}$

8. Factories and other industrial users account for about $\frac{23}{50}$ of the total water usage in the United States. Which of the following show that amount as a percent and decimal?
   F 46% and 0.46
   G 23% and 0.23
   H 50% and 0.5
   J 46% and 4.6

Holt Mathematics

Name _____ Date _____ Class _____

Problem Solving
Percent Problems

In 2000, the population of the United States was about 280 million people.
Use this information to answer each question.

1. About 20% of the total United States population is 14 years old or younger. How many people is that?

_____

2. About 6% of the total United States population is 75 years old or older. How many people is that?

_____

3. About 50% of Americans live in states that border the Atlantic or Pacific Ocean. How many people is that?

_____

4. About 12% of all Americans live in California. What is the population of California?

_____

5. About 7.5% of all Americans live in the New York City metropolitan area. What is the population of that region?

_____

6. About 12.3% of all Americans have Hispanic ancestors. What is the Hispanic American population here?

_____

Circle the letter of the correct answer.

7. Males make up about 49% of the total population of the United States. How many males live here?

   A 1,372 million   **C** 13.72 million

   B 137.2 million   **D** 1.372 million

8. About 75% of all Americans live in urban areas. How many Americans live in or near large cities?

   F 70 milliom   **H** 210 million

   G 200 million   **J** 420 million

9. About 7.4% of all Americans live in Texas. What is the population of Texas?

   A 74 million   **C** 7.4 million

   B 20.72 million   **D** 2.072 million

10. Between 1990 and 2000, the population of the United States grew by about 12%. What was the U.S. population in 1990?

   F 250 million   **H** 313.6 million

   G 33.6 million   **J** 268 million

Holt Mathematics

Name _____ Date _____ Class _____

Problem Solving
Using Percents

Use the table to answer each question.

Federal Income Tax Rates, 2001

Single Income	Tax Rate	Married Joint Income	Tax Rate
$0 to $27,050	15%	$0 to $45,200	15%
$27,051 to $65,550	27.5%	$45,201 to $109,250	27.5%
$65,551 to $136,740	30.5%	$109,251 to $166,500	30.5%
$136,741 to $297,350	35.5%	$166,501 to $297,350	35.5%
More than $297,350	39.1%	More than $297,350	31.5%

1. If a single person makes $25,000 a year, how much federal income tax will he or she have to pay?

2. If a married couple makes $148,000 together, how much federal income tax will they have to pay?

_____

3. The average salary for a public school teacher in the United States is $42,898. If two teachers are married, what is the average amount of federal income taxes they have to pay together?

4. In 2002 President George W. Bush received an annual salary of $400,000. Vice President Dick Cheney got $186,300. How much federal income tax do they each have to pay on their salary if they are married and filing jointly?

_____

Circle the letter of the correct answer.

5. Members of the U.S. Congress each earn $145,100 a year. How much federal income tax does each pay on their salary?

   A $51,510.50   C $21,765

   B $44,255.50   D $39,902.50

7. The average American with a college degree earns $33,365 a year. About how much federal income tax does he or she have to pay at a single rate?

   A $5,004.75   C $10,176.33

   B $9,175.38   D $11,844.58

6. A married couple each working a minimum-wage job will earn an average of $21,424 together a year. How much income tax will they pay?

   F $5,891.60   H $321.36

   G $3,213.60   J $6,534.32

8. The governor of New York makes $179,000 a year. How much federal income tax does that governor have to pay at a single rate?

   F $63,545   H $49,225

   G $54,595   J $26,850

Holt Mathematics

Name _____ Date _____ Class _____

Problem Solving
Building Blocks of Geometry

Place your hand down flat on a sheet of paper. Draw a point at the tip of your thumb, the tip of your middle finger, and the tip of your pinky.

1. Label the thumb point *A*, the middle finger point *B*, and the pinky point *C*.

    _____

2. Name all the planes you possibly can with points *A*, *B*, and *C*.

    _____

3. Draw and name all the lines you can make with points *A*, *B*, and *C*.

    _____

4. Name all the line segments possible using points *A*, *B*, and *C*.

    _____

5. Name all the rays possible using points *A*, *B*, and *C*.

    _____

    _____

6. Choose one line that you drew. Give all the different possible names for that line.

    _____

    _____

Circle the letter of the correct answer.

7. Which of the following has exactly one endpoint?

    A $\overleftrightarrow{OP}$

    B $\overline{AB}$

    C $\overleftrightarrow{TR}$

    D $\overrightarrow{SM}$

8. Which of the following is a straight path that extends without end in opposite directions?

    F a point

    G a line

    H a ray

    J a line segment

9. Which statement is false?

    A An infinite number of lines can be drawn through one point.

    B Exactly one line can be drawn between two points.

    C A line contains exactly one ray.

    D If points *A* and *B* are on a line, then line segment *AB* and line segment *BA* are the same.

10. Why is the false statement in Exercise 9 not true?

    F Any point on a line defines another ray on the line.

    G A line contains exactly two rays.

    H A line contains exactly five rays.

    J A line does not contain any rays.

Holt Mathematics

LESSON 8-2

Problem Solving

Measuring and Classifying Angles

Write the correct answer.

1. When a patient is lying flat in a hospital bed, what type of angle does the patient's body form? What is the measurement of that angle?

2. When a patient is sitting straight up in a hospital bed, the upper body has been raised to what angle? What type of angle is that?

3. Most hospital beds have a setting for the Fowler position. In this position, the patient's upper body is raised to form a 60° to 70° angle from a flat position. What types of angles are these?

4. What are the greatest and least differences between the straight-up position and the Fowler position in a hospital bed?

Circle the letter of the correct answer.

5. Medical technicians often set the handles of crutches so that the patient's elbow is at a 30° angle. What type of angle is this?

   A acute angle

   B right angle

   C obtuse angle

   D straight angle

6. By law, wheelchair ramps in public places cannot be greater than 5 degrees. Which type of angle does a wheelchair ramp in public form with the ground?

   F acute angle

   G right angle

   H obtuse angle

   J straight angle

7. Physical therapists use a goniometer to measure the extension of a sitting patient's knee. Resting is 90°, and full extension is 180°. What angle does the goniometer measure if the patient's knee is at $\frac{1}{2}$ extension?

   A 45°

   B 90°

   C 135°

   D 0°

8. The Q-angle is measured between two points on a patient's hip joint and one point on the knee joint. A normal Q-measure for men is 14° plus or minus 3 degrees. What type of angle is any normal Q-angle for men?

   F straight

   G obtuse

   H right

   J acute

Holt Mathematics

Name _____ Date _____ Class _____

Problem Solving
Angle Relationships

Use the two compass roses to answer questions 1-6.

Cardinal Directions

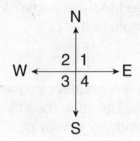

Intermediate Directions

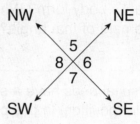

1. Which angles formed by the cardinal directions are vertical angles?

_____

2. Which angles formed by the intermediate directions are vertical angles?

_____

3. Draw the northwest directional ray on the cardinal compass rose. Describe the adjacent angles formed by the new ray.

_____

4. North on a compass is 0°, and east is 90°. Use this information to label the degrees for each direction on the two compass roses above.

Circle the letter of the correct answer.

5. Which angles formed by the cardinal directions are supplementary to ∠2?

  A ∠1

  B ∠1 and ∠3

  C ∠3 and ∠4

  D ∠1, ∠3 and ∠4

6. Which angles formed by the intermediate directions are supplementary to ∠6?

  F ∠5

  G ∠5 and ∠7

  H ∠7 and ∠8

  J ∠5, ∠7 and ∠8

7. Angles *A* and *B* are complementary. ∠*B* is twice as large as ∠*A*. What are the measurements for each angle?

  A ∠*A* = 45°; ∠*B* = 90°

  B ∠*A* = 30°; ∠*B* = 60°

  C ∠*A* = 60°; ∠*B* = 120°

  D ∠*A* = 90°; ∠*B* = 180°

8. ∠1 and ∠2 are complementary. ∠2 and ∠3 are supplementary. The measure of ∠1 is 45°. What is the measure of ∠3?

  F 45°

  G 270°

  H 90°

  J 135°

Holt Mathematics

Problem Solving
LESSON 8-4 · Classifying Lines

Use the map to answer each question.

1. The area where the borders of Utah, Colorado, Arizona, and New Mexico meet is sometimes called the Four Corners. What kind of lines are formed where the borders meet?

   _____

2. Which borderlines on the map are skew lines?

   _____

3. What kinds of lines are suggested by the eastern and western borders of New Mexico?

   _____

Western U.S. States

Circle the letter of the correct answer.

4. Which three states' borderlines intersect near the Grand Canyon?

   A Utah, Arizona, and Idaho

   B Idaho, Arizona, and Oregon

   C Nevada, Utah, and Arizona

   D Utah, Wyoming, and Idaho

5. Which two western states seem to have congruent borderlines?

   F Colorado and Wyoming

   G Oregon and Nevada

   H New Mexico and Nevada

   J Utah and Idaho

6. Which of the following do not appear to be parallel to the western borderline of Nevada?

   A the western borderline of California

   B the western borderline of Wyoming

   C the eastern borderline of Montana

   D the eastern borderline of Arizona

7. Which of these western states do not have borderlines that intersect near Great Salt Lake?

   F Utah and Nevada

   G Utah and Idaho

   H Utah and Wyoming

   J Utah and Colorado

Holt Mathematics

Name _____ Date _____ Class _____

Problem Solving
Triangles

Use the triangle diagram to answer each question.

1. Classify triangle *ABC*. What is the measure of the missing angle?

   _____

2. Classify triangle *XYZ*. What is the measure of the missing angle?

   _____

3. If triangle *MNO* is an equilateral triangle, what is the measure of the missing side?

   _____

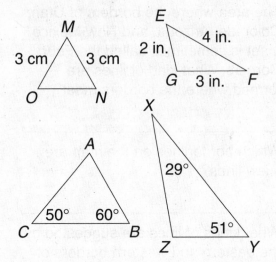

Circle the letter of the correct answer.

4. What is the complement of ∠*XYZ*?

   A 39°

   B 51°

   C 129°

   D 309°

5. Classify triangle *EFG*.

   F scalene triangle

   G isosceles triangle

   H equilateral triangle

   J right triangle

6. Which of the following statements is always true?

   A A right triangle is a scalene triangle.

   B An equilateral triangle is an isosceles triangle.

   C An isosceles triangle is an obtuse triangle.

   D A right triangle is an acute triangle.

7. Which of the following is not true of all right triangles?

   F The sum of the measures of the angles is 180°.

   G Two of its angles are supplementary angles.

   H At least two of its angles are acute.

   J The side with the greatest length is opposite the right angle.

Holt Mathematics

LESSON 8-6 **Problem Solving**
Quadrilaterals

Write the correct answer.

1. Fill in this Venn diagram using the terms quadrilaterals, squares, rectangles, rhombuses, parallelograms, and trapezoids.

2. Part of this quadrilateral is hidden. What could it possibly be?

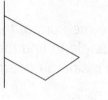

_____

_____

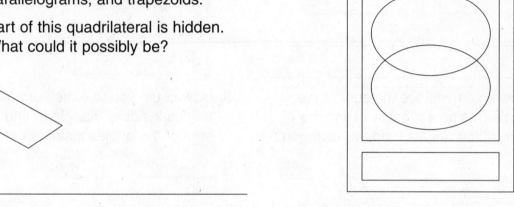

3. How could you make a trapezoid from a rectangle using only one cut?

_____

_____

_____

4. An engineer wants to build a building with a parallelogram base. He wants the four corners to be right angles and the four sides congruent. What type of base does the engineer want?

_____

_____

Circle the letter of the correct answer.

5. Each side of a quadrilateral-shaped picture frame has the same length. Which of the following is not a possible shape for the frame?

   A a rhombus

   B a square

   C a trapezoid

   D a parallelogram

6. The total length of the four sides of the picture frame from Exercise 5 is 4 feet, 8 inches. What is the length of each of its sides?

   F 14 inches

   G 1 foot, 3 inches

   H 12 inches

   J 2 inches

Holt Mathematics

Name _____ Date _____ Class _____

Problem Solving
Polygons

Write the correct answer.

1. Name each polygon in this figure.

   _____

   _____

   _____

   _____

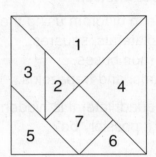

2. How could you use the sum of the angles inside a triangle to find the sum of the angles inside a heptagon?

   _____

   _____

   _____

   _____

3. How could you use the sum of the angles inside a triangle to find the sum of the angles inside a decagon?

   _____

   _____

   _____

   _____

4. In the space below, draw a rectangle and a parallelogram with side lengths congruent to the rectangle's. Now draw the diagonals for each of those polygons. What new polygons are formed by the diagonals in each quadrilateral?

5. In Exercise 4, what is true of the diagonals in the rectangle that isn't true of the diagonals of the parallelogram?

   _____

   _____

Circle the letter of the correct answer.

6. The perimeter of a regular hexagon is $13\frac{1}{2}$ inches. What is the length of each side?

   A $2\frac{7}{10}$ inches   **C** $3\frac{3}{4}$ inches

   B $2\frac{1}{4}$ inches   **D** $1\frac{11}{16}$ inches

7. Which of the following statements is sometimes false?

   F A plane figure is a polygon.

   G Each side of a polygon intersects exactly two other sides.

   H A polygon is a closed figure.

   J A polygon has straight sides.

Holt Mathematics

Name _____ Date _____ Class _____

LESSON 8-8 Problem Solving
Geometric Patterns

Complete this chart and look for patterns. Then answer the questions.

Number of Points on the Line	Draw and Label the Line and Points	Number of Different Line Segments in the Line
2	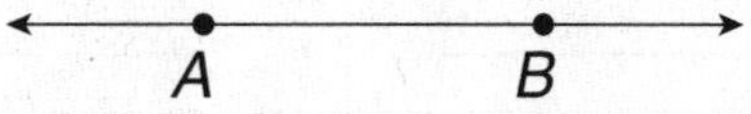 A B	1
1. 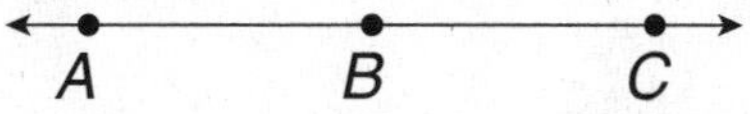	A B C	
2. 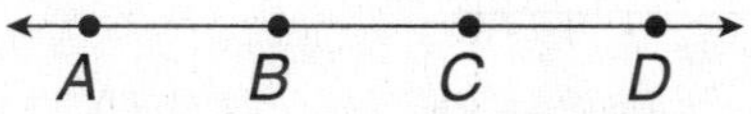	A B C D	
3. 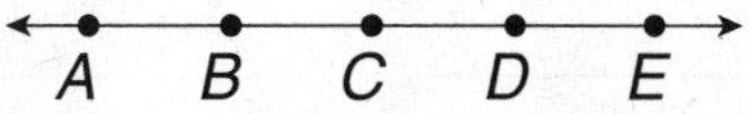	A B C D E	
4. 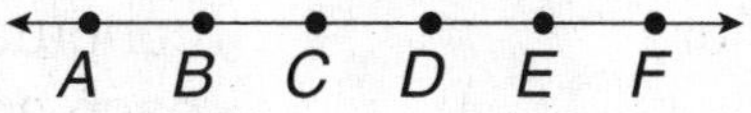	A B C D E F	

Circle the letter of the correct answer.

5. If n = the number of points on a line, which of the following expressions shows the number of different line segments on that line?

   A $2n - 3$

   B $(n^2 - n) \div 2$

   C $(n \div 2) \cdot 5$

   D $10n \div 2$

6. Using the pattern in the table and your answer to Exercise 5, how many different line segments will be on a line if there are 10 points on the line?

   F 17 line segments

   G 25 line segments

   H 45 line segments

   J 50 line segments

Holt Mathematics

LESSON 8-9 **Problem Solving**
Congruence

Write the correct answer.

1. Similar figures have the same shape but may have different sizes. How are similar figures different from congruent figures?

_____

_____

2. Pentagon A and Pentagon B are congruent regular polygons. If the total length of the sides of Pentagon B is 68.5 feet, what is the length of each side of Pentagon A?

_____

3. Is the following statement always true, sometimes true, or never true? Two congruent figures are similar figures. Explain.

_____

_____

_____

_____

4. Draw a figure congruent to this line segment. Explain how you drew your congruent figure.

•———————————•
A　　　　　　B

_____

_____

Circle the letter of the correct answer.

5. Which word makes this statement true? Corresponding parts of congruent figures are _____.

A not regular

B congruent

C polygons

D horizontal

6. If two angles of a right triangle are congruent, what are the measures of each angle in the triangle?

F 35°, 55°, and 90°

G 45°, 45°, and 90°

H 50°, 50°, and 90°

J 55°, 55°, and 90°

7. Which of the following polygons do not always have all congruent sides?

A a square

B an equilateral triangle

C a rhombus

D a pentagon

8. If ∠A of rectangle $ABCD$ is congruent to ∠X of triangle XYZ, which of these statements is true?

F Rectangle $ABCD$ is also a square.

G Triangle XYZ is a right triangle.

H Rectangle $ABCD$ is a regular polygon.

J Triangle XYZ is an acute triangle.

Holt Mathematics

Name _____ Date _____ Class _____

Problem Solving
Transformations

Write the correct answer.

1. If the rotation point of a circle is its center, how will all rotations affect the circle?

_____

2. What transformation could make an arrow pointing east become an arrow pointing north?

_____

3. What transformation could make the number 9 become the number 6?

_____

4. What transformation could make the letter P look like the letter b?

_____

5. On the coordinate plane at right, graph Triangle A with vertices (3, 1), (6, 1), and (3, 5). Then graph Triangle B with vertices (3, 6), (6, 6), and (3, 10). What transformation best describes the change from Triangle A to Triangle B?

_____

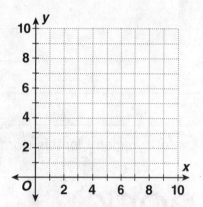

Circle the letter of the correct answer.

6. Which transformation best describes the figure on the right?

⌐ ⌐

A 90° clockwise rotation
B horizontal reflection
C 90° counterclockwise rotation
D horizontal translation

7. Which transformation best describes the figure on the left?

Ƨ Z

F horizontal reflection
G 180° counterclockwise rotation
H 90° counterclockwise rotation
J horizontal translation

Holt Mathematics

Name _____ Date _____ Class _____

Problem Solving
Line Symmetry

Write the correct answer.

1. Do your body and face appear to have a vertical line of symmetry or a horizontal line of symmetry?

   _____

2. Which letter of the alphabet has an infinite, or endless, number of lines of symmetry?

   _____

3. Ted says the diagonals of a rectangle are also its lines of symmetry. Do you agree? Explain.

   _____

   _____

   _____

4. Using the digits 0 through 9 and not repeating any digits, write a 3-digit number that has a horizontal line of symmetry.

   _____

   _____

   _____

5. Draw a line of symmetry for this word.

DOCK

6. Draw the lines of symmetry for this star.

Circle the letter of the correct answer.

7. How many lines of symmetry does this hexagon have?

   A 4

   B 8

   C 6

   D 2

8. How many lines of symmetry does this flower have?

   F 3

   G 4

   H 5

   J 6

9. How many lines of symmetry does a square have?

   A 0

   B 2

   C 4

   D 6

10. How many lines of symmetry does a regular pentagon have?

   F 1

   G 2

   H 4

   J 5

Holt Mathematics

LESSON 9-1 Problem Solving
Understanding Customary Units of Measure

Use customary units of measure to answer each question.

1. Which unit of measure would be most appropriate to use for the capacity of a swimming pool?

   _____

2. Which unit of measure would be most appropriate to use for the length of an insect?

   _____

3. Which unit of measure would be most appropriate to use for the weight of a television set?

   _____

4. Which unit of measure would be most appropriate to use for the weight of a feather?

   _____

5. Which unit of measure would be most appropriate to use for the distance between two cities?

   _____

6. Which unit of measure would be most appropriate to use for the capacity of a can of soup?

   _____

Circle the letter of the correct answer.

7. How long is a desk?
   A about 4 in.
   B about 4 ft
   C about 4 yd
   D about 4 mi

8. How much does a bird weigh?
   F about 3 oz
   G about 3 lb
   H about 3 T
   J about 30 T

9. How much does a can of soda hold?
   A about 1 glass of juice
   B about 4 small bottles of salad dressing
   C about 8 large containers of milk
   D about 10 spoonfuls

10. How long is your math book?
    F about 3 times the distance from your shoulder to your elbow
    G about 5 times the width of a classroom door
    H about 8 times the total length of 18 football fields
    J about 12 times the width of your thumb

Holt Mathematics

Name _____ Date _____ Class _____

Problem Solving
Understanding Metric Units of Measure

Use metric units of measure to answer each question.

1. Which unit of measure would be most appropriate to use for the capacity of a swimming pool?

   _____

2. Which unit of measure would be most appropriate to use for the length of an insect?

   _____

3. Which unit of measure would be most appropriate to use for the weight of a television set?

   _____

4. Which unit of measure would be most appropriate to use for the weight of a feather?

   _____

5. Which unit of measure would be most appropriate to use for the distance between two cities?

   _____

6. Which unit of measure would be most appropriate to use for the capacity of a can of soup?

   _____

Circle the letter of the correct answer.

7. How long is a desk?
   A about 1.5 mm
   B about 1.5 cm
   C about 1.5 m
   D about 1.5 km

8. What is the mass of a bird?
   F about 9 mg
   G about 90 mg
   H about 90 g
   J about 90 kg

9. What is the capacity of a can of soda?
   A about 5 mL
   B about 500 mL
   C about 5 L
   D about 500 L

10. How long is your math book?
   F about 30 times the width of a fingernail
   G about 10 times as thick as a dime
   H about 5 times as wide as a single bed
   J about 2 times the distance around a city block

Holt Mathematics

Problem Solving
Converting Customary Units

Write the correct answer.

1. Each side of a professional baseball base must measure 15 inches. What is the base's side length in feet?

2. In the NBA, any shot made from 22 feet or more from the basket is worth 3 points. How many yards from the basket is that?

_____

3. The maximum weight for a professional bowling ball is 16 pounds. What is the maximum weight in ounces?

4. A professional hockey goal is 6 feet wide and 4 feet high. What is the area of the goal in square yards?

_____

5. An NFL football field is 120 yards long. How many times would you have to run across the field to run 1 mile?

6. The official length for a marathon race is 26.2 miles. How many yards long is a marathon? How many feet?

_____

Circle the letter of the correct answer.

7. The distance between bases in a professional baseball game is 90 feet. What is the distance between bases in inches?

   A 1,000 inches   C 1,100 inches

   B 1,080 inches   D 10,800 inches

8. What is the area of a baseball diamond in square yards?

   F 300 square yards
   G 600 square yards
   H 900 square yards
   J 8,100 square yards

9. An NFL football can be no less than $\frac{87}{96}$ feet long. What is the minimum length for an official football in inches?

   A $10\frac{7}{8}$ inches   C $\frac{87}{1152}$ inches

   B $1\frac{3}{32}$ inches   D $2\frac{69}{96}$ inches

10. An official Olympic-sized swimming pool holds 880,000 gallons of water! How many fluid ounces of water is that?

   F 1,4080,000 fluid ounces
   G 7,040,000 fluid ounces
   H 112,640,000 fluid ounces
   J 1,760,000 fluid ounces

Holt Mathematics

Name _____ Date _____ Class _____

Problem Solving
Converting Metric Units

Write the correct answer.

1. The St. Gotthard Tunnel in Switzerland is the world's longest tunnel. It is 16.3 kilometers long. What is the tunnel's length in meters?

_____

2. Ostriches are the world's heaviest birds. On average, they weigh 156,500 grams. How many kilograms does the average ostrich weigh?

_____

3. The huge flower of the titam arum plant of Sumatra only lives for one day. During that time it grows 75 millimeters. What is the flower's height in centimeters?

_____

4. The average male elephant drinks about 120,000 milliliters of water each day. How many liters of water do most male elephants drink each day?

_____

Circle the letter of the correct answer.

5. The first successful steam locomotive pulled10,886.4 kilograms of iron. How many grams of iron did the locomotive pull?

   A 10.89 grams
   B 108.86 grams
   C 10,886,400 grams
   D 108,864,000 grams

6. The track used by the first successful steam locomotive was 15.3 kilometers long. How many meters long was the track?

   F 0.153 meter
   G 1.53 meters
   H 153 meters
   J 15,300 meters

7. About 2.03 meters of rain fall each year in a tropical rain forest. About how many centimeters of rainfall are there each year in a tropical rain forest?

   A 20.3 centimeters
   B 203 centimeters
   C 2,030 centimeters
   D 20,300 centimeters

8. The top layer of trees in a tropical forest has trees that can reach 6,096 centimeters in height. How many meters tall are these trees?

   F 6.096 meters
   G 60.96 meters
   H 609.6 meters
   J 609,600 meters

Holt Mathematics

Name _____ Date _____ Class _____

Problem Solving
Time and Temperature

Use the schedule to answer the questions.

1. Which bus from New York to Atlantic City would you take to spend the least amount of time on the bus?

   _____

2. Which bus would you take to spend the greatest amount of time on the bus?

   _____

3. Bus 231 took the same amount of time as Bus 230 to travel from New York to Atlantic City. If bus 231 left New York at 7:10 P.M., at what time did it arrive in Atlantic City?

   _____

New York to Atlantic City Schedule		
Bus	**Depart**	**Arrive**
225	7:30 A.M.	10:00 A.M.
226	9:50 A.M.	12:10 P.M.
227	11:00 A.M.	1:35 P.M.
228	1:45 P.M.	4:40 P.M.
229	3:10 P.M.	5:40 P.M.
230	6:00 P.M.	8:35 P.M.

Circle the letter of the correct answer.

4. Which measure is equivalent to 2 weeks?

   A 10 days

   B 336 hours

   C 2,016 minutes

   D 120,000 seconds

5. Which measure is NOT equivalent to the others?

   F $\frac{1}{4}$ day

   G 6 hours

   H 350 minutes

   J 21,600 seconds

6. Which is the best estimate?

   A 36°F is about 30°C.

   B 36°F is about 24°C.

   C 36°F is about 13°C.

   D 36°F is about 3°C.

7. Which is the best estimate?

   F 18°C is about 36°F.

   G 11°C is about 20°F.

   H 8°C is about 46°F.

   G 3°C is about 0°F.

Holt Mathematics

Name _____ Date _____ Class _____

Problem Solving
Finding Angle Measures in Polygons

Write the correct answer.

1. Most of the windows in a building are in the shape of a rectangle. What is the measure of one angle in each of those windows? What type of angle is it?

_____

2. The Pentagon Building in Washington, D.C. is in the shape of a regular pentagon. What is the measure of one angle in the Pentagon Building? What type of angle is it?

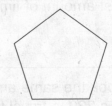

_____

3. Most cells in a honeycomb are in the shape of a regular hexagon. What is the measure of one angle in each of those cells? What type of angle is it?

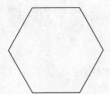

_____

4. Most sports pennants are in the shape of an isosceles triangle. What is the measure of the smaller angle in this sports pennant? What type of angle is it?

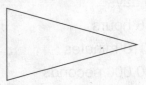

_____

Circle the letter of the correct answer.

5. What is the measure of a corner of a square piece of note paper?

   A 45°

   B 90°

   C 145°

   D 180°

6. What type of angle is the corner of a square piece of note paper?

   F acute angle

   G right angle

   H obtuse angle

   J straight angle

Holt Mathematics

Problem Solving

LESSON 9-7 *Perimeter*

Write the correct answer.

1. Use a ruler to find the perimeter of your math textbook in inches.

_____

_____

2. Use a ruler to find the perimeter of your desk in feet and inches.

_____

_____

3. The world's largest flag weighs 3,000 pounds and requires at least 500 people to set up! This United States flag is 505 feet long and 255 feet wide. What is the perimeter of this United States flag?

_____

4. Students in Lisbon, Ohio, built the world's largest mousetrap in 1998. The mousetrap is 9 feet 10 inches long and 4 feet 5 inches wide—and it actually works! What is the perimeter of the mousetrap in feet and inches?

_____

Circle the letter of the correct answer.

5. The giant ball dropped every New Year's Eve in New York City is covered with 504 crystal equilateral triangles. The average perimeter of each triangle is $15\frac{3}{4}$ inches. What is the average side length of each crystal triangle on the ball?

   A 5 inches

   B $5\frac{1}{8}$ inch

   C $5\frac{1}{4}$ inch

   D $5\frac{1}{2}$ inch

6. United States dollar bills are 2.61 inches wide and 6.14 inches long. Larger notes in circulation before 1919 measured 3.125 inches wide by 7.4218 inches long. What is the difference between the old and new dollar bill perimeters?

   F 3.5936 inches

   G 3.9536 inches

   H 4.0956 inches

   J 4.5936 inches

7. The perimeter of regular octagon-shaped swimming pool is 42 feet. What is the length of each side of the pool?

   A 5 feet

   B 5 feet 3 inches

   C 5 feet 2 inches

   D 5.2 feet

8. Each Scrabble® tile is 1.8 centimeters wide and 2.1 centimeters tall. If the tiles spell the word LOVE, what is the perimeter of the entire word?

   F 7.8 cm

   G 18.6 cm

   H 12 cm

   J 31.2 cm

Holt Mathematics

LESSON 9-8	**Problem Solving**

Circles and Circumference

Use the table to answer each question. Use 3.14 for π.

1. Which coin has the smallest radius?
How long is that coin's radius?

_____

2. What is the circumference of a
nickel?

_____

3. What is the circumference of a
quarter?

_____

4. Which coin has a greater
circumference, a dollar or half dollar?
What is the difference in their
circumferences?

_____

Official U.S. Coin Sizes

Coin	**Diameter** (rounded to nearest mm)
Penny	19
Nickel	21
Dime	18
Quarter	24
Half Dollar	31
Dollar	27

5. If you rolled a dollar coin on its edge,
how far would it go with each
complete turn?

_____

6. Which U.S. coins will fit in a
vending machine coin slot that is
2 centimeters wide?

_____

Circle the letter of the correct answer.

7. A dime has 118 ridges evenly spaced
along its circumference. About how
wide is each ridge?

A about 0.24 mm

B about 0.48 mm

C about 0.15 mm

D about 0.08 mm

8. The engraved words "United States
of America" run about one-half the
circumference of all U.S. coins. On
which coin will the words run about
38 mm?

F penny   **H** quarter

G dime   **J** half dollar

9. You have two coins with a total
circumference of 116.18 mm. How
much money do you have?

A $0.02   **C** $0.11

B $0.06   **D** $0.35

10. You have three coins with a total
circumference of 216.66 mm. How
much money do you have?

F $0.15   **H** $0.30

G $0.25   **J** $0.55

Holt Mathematics

LESSON 10-1 Problem Solving
Estimating and Finding Area

Use the table to answer each question.

State Information

State	Approx. Width (mi)	Approx. length (mi)	Water Area (mi²)
Colorado	280	380	376
Kansas	210	400	462
New Mexico	343	370	234
North Dakota	211	340	1,724
Pennsylvania	160	283	1,239

1. New Mexico is the 5th largest state in the United States. What is its approximate total area?

2. Kansas is the 15th largest state in the United States. What is its approximate total area?

3. What is the difference between North Dakota's land area and water area?

4. What is Pennsylvania's approximate land area?

Circle the letter of the correct answer.

5. What is the difference between Colorado's land area and Pennsylvania's land area?
   A 106,400 mi²
   B 61,120 mi²
   C 60,120 mi²
   D 45,280 mi²

6. About what percent of the total area of Pennsylvania is covered by land?
   F about 3%
   G about 30%
   H about 67%
   J about 97%

7. Rhode Island is the smallest state. Its total land area is approximately 1,200 mi². Rhode Island is approximately 40 miles long. About how wide is Rhode Island?
   A about 20 mi
   B about 40 mi
   C about 50 mi
   D about 30 mi

8. The entire United States covers 3,794,085 square miles of North America. About how much of that area is not made up of the 5 states in the chart?
   F 2,537,470 mi²
   G 3,359,755 mi²
   H 3,686,525 mi²
   J 3,1310,818 mi²

Holt Mathematics

Problem Solving

Area of Triangles and Trapezoids

Use the quilt design to answer the questions.

1. What are the lengths of the bases of each trapezoid?

   _____

2. What is the height of each trapezoid?

   _____

3. What is the area of each trapezoid?

   _____

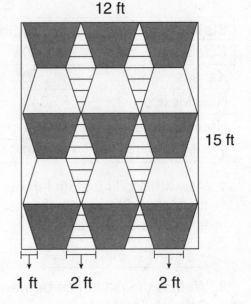

12 ft

15 ft

1 ft   2 ft   2 ft

Circle the letter of the correct answer.

4. What is the length of the base of each striped triangle?

   A 1 ft

   B 2 ft

   C 3 ft

   D 4 ft

5. What is the height of each striped triangle?

   F 1 ft

   G 2 ft

   H 3 ft

   J 5 ft

6. What is the area of each striped triangle?

   A 3 ft²

   B 1 ft²

   C $\frac{3}{4}$ ft²

   D $\frac{1}{4}$ ft²

7. What is the area of the quilt?

   F 36 ft²

   G 90 ft²

   H 96 ft²

   J 180 ft²

Holt Mathematics

Name _____ Date _____ Class _____

Write the correct answer.

1. The shape of Nevada can almost be divided into a perfect rectangle and a perfect triangle. About how many square miles does Nevada cover?

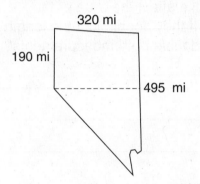

2. The shape of Oklahoma can almost be divided into 2 perfect rectangles and 1 triangle. About how many square miles does Oklahoma cover?

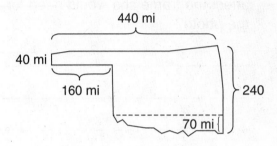

_____

3. The front side of an apartment building is a rectangle 60 feet tall and 25 feet wide. Bricks cover its surface, except for a door and 10 windows. The door is 7 feet tall and 3 feet wide. Each window is 4 feet tall and 2 feet wide. How many square feet of bricks cover the front side of the building?

4. Each side of a square garden is 12 meters long. A hedge wall 1 meter wide surrounds the garden. What is the area of the entire garden including the hedge wall? How many square meters of land does the hedge wall cover alone?

_____

Circle the letter of the correct answer.

5. A figure is formed by a square and a triangle. Its total area is 32.5 m². The area of the triangle is 7.5 m². What is the length of each side of the square?

   A 5 meters   C 15 meters

   B 25 meters   D 16.25 meters

6. A rectangle is formed by two congruent right triangles. The area of each triangle is 6 in². If each side of the rectangle is a whole number of inches, which of these could not be its perimeter?

   F 26 inches   H 24 inches

   G 16 inches   J 14 inches

Holt Mathematics

Problem Solving

LESSON 10-4 *Comparing Perimeter and Area*

Write the correct answer.

1. Fiona's school photograph is 6 inches long and 5 inches wide. If she orders a triple enlargement how would this affect the area of the photo? How would the enlargement affect the frame she would need for the photo?

_____

_____

_____

_____

2. The Whitman's kitchen is 8 feet long and 6 feet wide. They are planning on renovating the kitchen to have more space. If they double just the width, how will it affect the area of the room? If they double just the length? If they double both measurements?

_____

_____

_____

_____

Circle the letter of the correct answer.

3. Kent saw a table in a magazine that was 3 feet wide and 4 feet long. If he wants to make a similar version of the table with an area 4 times larger, what dimensions should he use?

   A 4 ft wide and 5 ft long

   B 6 ft wide and 8 ft long

   C 9 ft wide and 12 ft long

   D 12 ft wide and 16 ft long

4. The triangular sail on Shakeera's boat is 8 meters wide and 10 meters tall. She wants to make a model of the boat that is $\frac{1}{20}$ of its actual size. How much canvas will Shakeera use for the model boat's sail?

   F 10 m^2 of canvas

   G 1 m^2 of canvas

   H 0.1 m^2 of canvas

   J 0.01 m^2 of canvas

5. A triangle is 6.4 cm long and 8.2 cm tall. If you triple its dimensions, what would be the area of the enlarged triangle?

   A 78.72 cm^2 **C** 236.16 cm^2

   B 157.44 cm^2 **D** 472.32 cm^2

6. The dimensions of a regular pentagon are doubled. The perimeter of the enlarged pentagon is 25 yards. What was the length of each side of the original pentagon?

   F 2.5 yards   **H** 5 yards

   G 12 yards   **J** 16.25 yards

Holt Mathematics

Name _____ Date _____ Class _____

Problem Solving
Area of Circles

Use the table to answer each question. Use 3.14 for *pi*.

1. Which ring is the largest? What area does it enclose?

_____

2. What is the area of the center circle, or the inner 10 scoring ring, on the target?

_____

3. What area does Ring 5 enclose?

_____

Official Archery Target Ring Diameters

Scoring Ring	Diameter (cm)
1	80
2	72
3	64
4	56
5	48
6	40
7	32
8	24
9	16
10	8
Inner 10	4

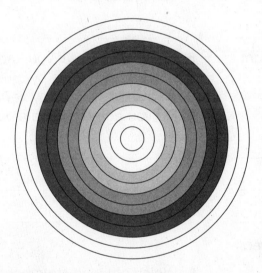

Circle the letter of the correct answer.

4. Which ring encloses an area of 4069.44 cm²?

A Ring 2

B Ring 3

C Ring 6

D Ring 8

5. How much greater is the area enclosed by Ring 10 than the area enclosed by Ring 9?

F 50.24 cm²

G 150.72 cm²

H 200.96 cm²

J 251.2 cm²

6. What is the area enclosed by Ring 6?

A 5,024 cm²

B 1,600 cm²

C 1,256 cm²

D 62.8 cm²

7. What is the area enclosed by Ring 1?

F 10 times the area of Ring 10

G 20 times the area of Ring 10

H 100 times the area of Ring 10

J 1,000 times the area of Ring 10

Holt Mathematics

Name _____ Date _____ Class _____

Problem Solving
Three-Dimensional Figures

Write the correct answer.

1. Pamela folded an origami figure that has 5 faces, 8 edges, and 5 vertices. What kind of three-dimensional figure could Pamela have created?

_____

_____

2. Look at your classroom chalkboard. What kind of three-dimensional figure is the board eraser? What kind of three-dimensional figure is the chalk?

_____

_____

3. If you cut a cylinder in half between its two bases, what two three-dimensional figures are formed?

_____

_____

4. You have two hexagons. How many rectangles do you need to create a hexagonal prism?

_____

_____

5. All four of the faces of a paperweight are triangles. Is this enough information to classify this three-dimensional figure? Explain.

_____

_____

_____

6. Paulo says that if you know the number of faces a pyramid has, you also know how many vertices it has. Do you agree? Explain.

_____

_____

_____

Circle the letter of the correct answer.

7. How is a triangular prism different from a triangular pyramid?

   A The prism has 2 bases.

   B The pyramid has 2 bases.

   C All of the prism's faces are triangles.

   D The pyramid has 5 faces.

8. Which of these statements is not true about a cylinder?

   F It has 2 circular bases.

   G It has a curved lateral surface.

   H It is a solid figure.

   J It is a polyhedron.

9. A museum needs to ship a sculpture that has a curved lateral surface and one flat circular base. In what shape box should they mail the sculpture?

   A cone   **C** cylinder

   B cube   **D** triangular prism

10. A glass prism reflects white light as a multicolored band of light called a spectrum. The prism has 5 glass faces with 9 edges and 6 vertices. What kind of prism it it?

    F cube   **H** triangular pyramid

    G cone   **J** triangular prism

Holt Mathematics

LESSON 10-7

Problem Solving
Volume of Prisms

Write the correct answer.

1. At 726 feet tall, Hoover Dam is one of the world's largest concrete dams. In fact, it holds enough concrete to pave a two-lane highway from New York City to San Francisco! The dam is shaped like a rectangular prism with a base 1,224 feet long and 660 feet wide. About how much concrete forms Hoover Dam?

2. The Vietnam Veterans Memorial in Washington, D.C., is a 493.5-foot-long wall made of polished black granite engraved with the names of soldiers who died in the war. The wall is 0.25 feet thick and has an average height of 9 feet. About how many cubic feet of black granite was used in the Vietnam Veterans Memorial?

3. Benitoite, a triangular prism crystal, is the official state gem of California. One benitoite crystal found in California is 1.2 cm tall, with a base width of 2 cm and a base height of 1.3 cm. How many cubic centimeters of benitoite are in that crystal?

4. The Flatiron Building in New York City is a triangular prism. A solid bronze souvenir model of the building is 5 inches tall, with a base height of 1.5 inches and a base width of 2.5 inches. How much bronze was used to make the model?

Circle the letter of the correct answer.

5. Individual slices of pizza are sold in 2-inch-tall triangular prism boxes. The box base is 8 inches wide, with a 7-inch height. How many cubic inches of pizza will fit in each box?

   A 112 in^3 **C** 60 in^3

   B 102 in^3 **D** 56 in^3

6. The world's largest chocolate bar is a huge rectangular prism weighing more than a ton! The bar is 9 feet long, 4 feet tall, and 1 foot wide. How many cubic feet of chocolate does it have?

   F 13 ft^3 **H** 36 ft^3

   G 14 ft^3 **J** 72 ft^3

7. A box can hold 175 cubic inches of cereal. If the box is 7 inches long and 2.5 inches wide, how tall is it?

   A 25 in.

   B 10 in.

   C 17.5 in.

   D 9.5 in.

8. A triangular prism used to reflect light is made of 120 cm^3 of glass. If the prism is 5 centimeters tall, what is the area of each of its triangular bases?

   F 24 cm

   G 12 cm

   H 12 cm^2

   J 24 cm^2

Holt Mathematics

Problem Solving
Volume of Cylinders

Write the correct answer.

1. The Hubble Space Telescope was launched into space in 1990. Shaped like a cylinder, the telescope is 15.9 meters long, with a diameter of 4.2 meters. To the nearest whole cubic meter, what is the volume of the Hubble Space Telescope?

2. The Living Color aquarium in Bermuda is the largest freestanding cylindrical aquarium in the Western Hemisphere. With a 10-foot diameter and an 18-foot height, the aquarium holds 10,400 gallons of water! What is the aquarium's volume in cubic feet?

3. In 1902 an American music company built the world's largest music recording cylinder. Nicknamed "Brutus," the cylinder is 5 feet tall, with a 2-foot diameter. What is the volume of the "Brutus" cylinder?

4. The world's largest glass of orange juice was filled in Florida in 1998. At 8 feet tall and with a 2-foot radius, the glass held about 700 gallons of orange juice. What was the volume of that huge glass of orange juice?

Circle the letter of the correct answer.

5. A large can of soda is 7.5 inches tall and has a 3-inch diameter. A small can of soda is 5 inches tall with a 2.5-inch diameter. To the nearest cubic inch, how much more soda does the large can hold?

   A 53 in^3 more soda

   B 28 in^3 more soda

   C 25 in^3 more soda

   D 20 in^3 more soda

6. A cylindrical candle is tightly packed in a rectangular box with a volume of 144 in^3. Which of these could be the dimensions of the candle?

   F $h = 6$ in.; $r = 3$ in.

   G $h = 2$ in.; $r = 5$ in.

   H $h = 4$ in.; $r = 3$ in.

   J $h = 3$ in.; $r = 4$ in.

7. The maximum length for an official professional baseball bat is 36 inches. Its maximum diameter is 2.6 inches. To the nearest cubic inch, what is the maximum volume of a professional baseball bat?

   A 21 in^3 **C** 191 in^3

   B 119 in^3 **D** 764 in^3

8. A can of tennis balls is 21 centimeters tall and has a diameter of 8 centimeters. What is the volume of the tennis ball can?

   F 17,408.16 cm^3 **H** 527.52 cm^3

   G 1,055.04 cm^3 **J** 263.76 cm^3

Holt Mathematics

LESSON 10-9 Problem Solving
Surface Area

Write the correct answer.

1. The world's largest cookie was baked in Wisconsin in 1992. Its diameter was 34 feet and contained about 4 million chocolate chips! If the cookie was a cylinder 1 foot tall, and you wanted to cover it with icing, how many square inches would you have to ice? Use 3.14 for π.

2. The top of the Washington Monument is a square pyramid covered with white marble. Each triangular face is 58 feet tall and 34 feet wide. About how many square feet of marble covers the top of the monument? (The base is hollow.)

3. The Parthenon, a famous temple in Greece, is surrounded by large stone columns. Each column is 10.4 meters tall and has a diameter of 1.9 meters. To the nearest whole square meter, what is the surface area of each column (not including the top and bottom)?

4. The tablet that the Statue of Liberty holds is 7.2 meters long, 4.1 meters wide, and 0.6 meters thick. The tablet is covered with thin copper sheeting. If the tablet was freestanding, how many square meters of copper covers the statue's tablet?

Circle the letter of the correct answer.

5. The largest Egyptian pyramid is called the Great Pyramid of Khufu. It has a 756-foot square base and a slant height of 481 feet. What is the total surface area of the faces of the Pyramid of Khufu?

   A 727,272 ft²   C 727,727 ft²

   B 727,722 ft²   D 772,272 ft²

7. A can of frozen orange juice is 7.5 inches tall, and its base diameter is 3.5 inches. What size strip of paper is used for its label?

   A 82.43 in²   C 576.98 in²

   B 26.25 in²   D 101.66 in²

6. A glass triangular prism for a telescope is 5.5 inches tall. Each side of the triangular base is 4 inches long, with a 3-inch height. How much glass covers the surface of the prism?

   F 6 in²   H 39 in²

   G 12 in²   J 78 in²

8. Tara made fuzzy cubes to hang in her car. Each side of the 2 cubes is 4 inches long. How much fuzzy material did Tara use to make both cubes?

   F 96 in²   H 16 in²

   G 192 in²   J 128 in²

Holt Mathematics

Name _____ Date _____ Class _____

LESSON 11-1

Problem Solving

Integers in Real-World Situations

Write the correct answer.

1. The element mercury is used in thermometers because it expands as it is heated. Mercury melts at 38°F below zero. Write this temperature as an integer.

2. Denver, Colorado, earned the nickname "Mile High City" because of its elevation of 5,280 feet above sea level. Write Denver's elevation as an integer in feet and miles.

3. The lowest temperature recorded in San Francisco was 20°F. Buffalo's lowest recorded temperature was the opposite of San Francisco's. What was Buffalo's record temperature?

4. Greenland holds the record for the lowest temperature recorded on Earth. That temperature in degrees Fahrenheit is 65 degrees below zero. What is Earth's lowest recorded temperature written as an integer?

5. In 1960, explorers on the submarine *Trieste 2* set the world record for the deepest dive. The ship reached 35,814 feet below sea level. Write this depth as an integer.

6. In 1960, Joseph W. Kittinger, Jr., set the record for the highest parachute jump. He jumped from an air balloon at 102,800 feet above sea level. Write this altitude as an integer.

Circle the letter of the correct answer.

7. Which situation cannot be represented by the integer −10?

   A an elevation of 10 feet below sea level

   B a temperature increase of 10°F

   C a golf score of 10 under par

   D a bank withdrawal of $10

8. Paper was invented in China one thousand, nine hundred years ago. Which integer represents this date?

   F 1,900

   G 900

   H −1,900

   J −1,000

9. The elevation of the Dead Sea is about 1,310 feet below sea level. Which integer represents this elevation?

   A −1,310

   B −131

   C 131

   D 1,310

10. The quarterback had a 10-yard loss and then a 25-yard gain. Which integer represents a 25-yard gain?

   F −25

   G −10

   H 25

   J 10

Holt Mathematics

Name _____ Date _____ Class _____

Problem Solving
Comparing and Ordering Integers

Use the table below to answer each question.

Continental Elevation Facts

Continent	Highest Point	Elevation (ft) above sea level	Lowest Point	Elevation (ft) below sea level
Africa	Mount Kilimanjaro	19,340	Lake Assal	−512
Antarctica	Vinson Massif	16,066	Bentley Subglacial Trench	−8,327
Asia	Mount Everest	29,035	Dead Sea	−1,349
Australia	Mount Kosciusko	7,310	Lake Eyre	−52
Europe	Mount Elbrus	18,510	Caspian Sea	−92
North America	Mount McKinley	20,320	Death Valley	−282
South America	Mount Aconcagua	22,834	Valdes Peninsula	−131

1. What is the highest point on Earth? What is its elevation?

_____

_____

2. What is the lowest point on Earth? What is its elevation?

_____

_____

3. Which point on Earth is higher, Mount Elbrus or Mount Kilimanjaro?

_____

4. Which point on Earth is lower, the Caspian Sea or Lake Eyre?

_____

Circle the letter of the correct answer.

5. Which continent has a higher elevation than North America?

A Antarctica

B South America

C Europe

D Australia

6. Which continent has a lower elevation than Africa?

F Australia

G Europe

H Asia

J South America

7. Write the continents in order by their highest points, from highest elevation to lowest elevation.

_____

_____

Holt Mathematics

Name _____ Date _____ Class _____

Problem Solving
The Coordinate Plane

Use the coordinate plane on the map of Texas below to answer each question.

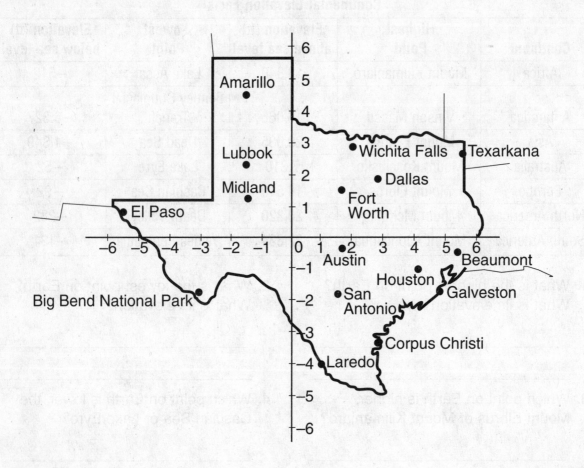

1. Which location in Texas is closest to the ordered pair (5, −2)?

   _____

2. What ordered pair best describes the location of Dallas, Texas?

   _____

3. Which location in Texas is closest to the ordered pair (−6, 1)?

   _____

4. Which location in Texas is located in Quadrant III of this coordinate plane?

   _____

5. Which three locations in Texas all have positive *y*-coordinates and nearly the same *x*-coordinate?

   _____

6. Which cities on this map of Texas have locations with *y*-coordinates less than −3?

   _____

Holt Mathematics

Problem Solving
LESSON 11-4 *Adding Integers*

In 1997, Tiger Woods became the youngest golfer ever to win the Masters Tournament. There are four rounds of 18 holes in the Masters Tournament. Use Woods's scorecard to answer questions 1–6.

Tiger Woods																		
Hole	1	2	3	4	5	6	7	8	9	10	11	12	13	14	15	16	17	18
Rd. 1	1	0	0	1	0	0	0	1	1	−1	0	−1	−1	0	−2	0	−1	0
Rd. 2	0	−1	1	0	−1	0	0	−1	0	0	0	0	−2	−1	−1	0	0	0
Rd. 3	0	−1	0	0	−1	0	−1	−1	0	0	−1	0	0	0	−1	0	0	−1
Rd. 4	0	−1	0	0	1	0	1	−1	0	0	−1	0	−1	−1	0	0	0	0

1. What was Woods's total score for round 1 of the tournament?

_____

2. What was his total score for the second round of the tournament?

_____

3. What was his total score for the third round of the tournament?

_____

4. What was his total score for the fourth round of the tournament?

_____

Circle the letter of the correct answer.

5. Woods's final score in 1997 was the lowest in the history of the Masters Tournament. What was Woods's record-breaking final score?

   A −16

   B −17

   C −18

   D −20

6. Tom Kite placed second in the 1997 Masters Tournament. His final score was 12 strokes higher than Tiger Woods's final score. What was Kite's final score?

   F −30

   G −12

   H − 8

   J 0

7. Which of the following is the sum of Woods's scores on the 8th hole?

   A 2

   B 1

   C −1

   D −2

8. Which of the following is the sum of Woods's scores on the 15th hole?

   F 4

   G −4

   H 0

   J 1

Holt Mathematics

Name _____ Date _____ Class _____

Problem Solving
Subtracting Integers

Write the correct answer.

1. The average surface temperature on Earth is 59°F. The average surface temperature on Mars is 126°F lower than on Earth. What is the average surface temperature on Mars?

_____

2. The average surface temperature on Saturn is 46°F colder than on Jupiter. Jupiter's average surface temperature is 162°F below zero. What is the average surface temperature on Saturn?

_____

3. Venus has the hottest average surface temperature at 854°F. Mercury, the planet closest to the Sun, has an average surface temperature that is 522°F colder than Venus's. What is Mercury's average surface temperature?

_____

4. Pluto has an average surface temperature of 355°F below zero. Neptune, its closest neighbor, has the coldest average surface temperature. It is 10°F colder on Neptune than on Pluto. What is the average surface temperature on Neptune?

_____

Circle the letter of the correct answer.

5. Which of the following is the difference between 247°F below zero and 221°F above zero?

    A −26°F

    B 129°F

    C −468°F

    D 468°F

6. Which of the following is the difference between 806°C above zero and 328°C below zero?

    F 1,134°C

    G 478°C

    H −478°C

    J −1,134°C

7. Which of the following is the difference between −40°C and −30°C?

    A −10°C

    B 708C

    C −120°C

    D −1°C

8. Which of the following is the difference between 8,700°F and −344°F?

    F 8,356°F

    G 900°F

    H −9°F

    J 9,044°F

Holt Mathematics

LESSON 11-6 Problem Solving
Multiplying Integers

Write the correct answer.

1. The coldest temperature ever recorded in Rhode Island was 25°F below zero. Though Nevada lies much farther south, its coldest temperature was twice as cold as Rhode Island's. What was Nevada's record cold temperature?

_____

2. Tom and Kim made up a game in which black tiles equal +5 points each, and red tiles equal −3 points each. The person with the most points wins. At the end of the game Tom had 6 red tiles and 4 black tiles, and Kim had 4 red tiles and 3 black tiles. Who won?

_____

3. During a month-long drought, the amount of water in the family's well changed −4 gallons a day. How much did the amount of water in the well change after one week?

_____

4. Sperm whales dive deeper than any other mammals. They regularly dive to 3,937 feet below sea level. But they sometimes dive to twice this depth! To what depth can sperm whales dive?

_____

Circle the letter of the correct answer.

5. On Monday morning, the value of LCM stock was $15 a share. Then the value of the stock changed by −3 dollars a day for 4 days in a row. What was the value of one share of LCM stock after the fourth day?

   A $1

   B $3

   C $6

   D $12

6. Lake Manitoba and Lake Winnipeg are two of the largest lakes in Canada. The greatest depth of Lake Manitoba is 12 feet. Lake Winnipeg is 5 times deeper than Lake Manitoba. What is the greatest depth of Lake Winnipeg?

   F 5 feet

   G 17 feet

   H 50 feet

   J 60 feet

7. Which addition expression could be used to check the product of $5 \cdot (-3)$?

   A $5 + 5 + 5$

   B $-3 + (-3) + (-3)$

   C $5 + 5 + 5 + 5 + 5$

   D $-3 + (-3) + (-3) + (-3) + (-3)$

8. Which property allows you to rewrite $-2 \cdot (-4)$ as $-4 \cdot (-2)$?

   F Commutative Property

   G Distributive Property

   H Integer Property

   J Associative Property

Holt Mathematics

LESSON **11-7** # Problem Solving

Dividing Integers

Use the table below to answer questions 1–6.

Temperatures for Barrow, Alaska

	JAN	FEB	MAR	APRIL	MAY	JUNE	JULY	AUG	SEPT	OCT	NOV	DEC
Temp (°F)	−13	−18	−15	−2	19	34	39	38	31	14	−2	−11

1. What is the average temperature in Barrow for December and January?

_____

2. What is the average temperature in Barrow for March and July?

_____

3. Which month's average temperature is half as warm as August's?

_____

4. What is the average temperature in Barrow for October and November?

_____

5. What is the average temperature in Barrow for January through April?

_____

6. What is the city's average temperature for September through December?

_____

Circle the letter of the correct answer.

7. A submarine dove to a depth of 168 feet in 7 minutes. What was the average rate of change in its location?

   A 24 feet

   B 168 feet

   C −24 feet

   D −168 feet

8. In its first 4 months of business, Skyscraper Records reported its losses as −$1,520. What was the company's average monthly loss?

   F −$1,520

   G −$380

   H −$38

   J $380

9. Which of these expressions checks the solution to the division problem $-8 \div (-2) = 4$?

   A $-8 \cdot (-2)$

   B $4 \cdot 4$

   C $-2 \cdot (2)$

   D $4 \cdot (-2)$

10. A glacier is melting 3 in^3 a year. At that rate, how long will it take for the glacier to change by −24 in^3?

   F 72 years

   G 6 years

   H 8 years

   J 24 years

Holt Mathematics

Name _____ Date _____ Class _____

Problem Solving
Solving Integer Equations

For questions 1–8, the temperatures found are in °F.

1. The highest recorded temperature in Africa is the solution to $x \div (-4) = -34$. What is Africa's highest recorded temperature?

2. The lowest recorded temperature in Australia is the solution to $7x = -56$. What is Australia's lowest recorded temperature?

_____

3. To find Africa's lowest recorded temperature, solve the following equation: $80 - x = 91$.

4. To find Europe's highest recorded temperature, solve the following equation: $x \div -2 = -61$.

_____

5. The solution to $-2x = -116$ is the highest recorded temperature in Antartica. What is Antartica's highest recorded temperature?

6. The solution to $x + (-23) = -90$ is the lowest recorded temperature in Europe. What is Europe's lowest recorded temperature?

_____

Circle the letter of the correct answer.

7. Which of the following is a solution to $x + (-11) = -140$?

   A 12

   B −129

   C −151

   D −1,540

8. Which of the following is a solution to $-110 + x = 19$?

   F 91

   G 129

   H −5

   J −2,090

9. Which of the following is a solution to $5x = -75$?

   A −375

   B −80

   C −70

   D −15

10. Which of the following is a solution to $-270 \div x = -30$?

   F 8,100

   G −300

   H 9

   J −240

Holt Mathematics

Name _____ Date _____ Class _____

Problem Solving
Tables and Functions

Use the tables to answer each question.

Table 1	
miles	kilometers
2	3.22
3	4.83
4	
5	8.05

Table 2	
ounces	grams
1	28.35
2	
3	85.05
4	113.4

Table 3	
gallons	liters
5	
10	37.9
15	56.85
20	75.8

1. Write an equation for a function that gives the values in table 1. Define the variables you use. Use your equation to find the missing term in the table.

_____

_____

_____

3. Write an equation for a function that gives the values in table 3. Define the variables you use. Use your equation to find the missing term in the table.

_____

_____

_____

variables you use. Use your equation to find the missing term in the table.

_____

_____

_____

4. There are 4 quarts in a gallon. Write an equation for a function relating quarts to liters. Then use your equation to find how many liters of oil a 50-quart barrel can hold.

_____

_____

_____

Circle the letter of the correct answer.

5. The Rocky Mountains stretch 3,750 miles across North America. What is this length in kilometers?

   A 2,329.2 kilometers

   B 1,164.6 kilometers

   C 6,037.5 kilometers

   D 12,075 kilometers

2. Write an equation for a function that gives the values in table 2. Define the

6. A hummingbird egg only weighs 0.25 grams! How many ounces does the egg weigh?

   F about 7.0875 ounces

   G about 0.009 ounces

   H about 28 ounces

   J about 9 ounces

Holt Mathematics

Name _____ Date _____ Class _____

LESSON	**Problem Solving**
11-10	*Graphing Functions*

Use the table to answer each question.

1. $F = \frac{9}{5}C + 32$ is an equation for the function that gives the values in the table. What does each variable represent in the equation? Use the equation to complete the table.

_____

_____

Equivalent Temperatures

Celsius (°C)	Fahrenheit (°F)
−20	−4
−10	14
0	32
10	
20	

2. Write a different equation for a function that gives the values in the table.

_____

_____

3. Is the ordered pair (30, 86) a solution for either equation? Why or why not? What does each value in the ordered pair represent?

_____

_____

4. Graph the function described by either equation on the graph at right.

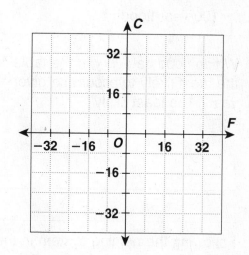

Circle the letter of the correct answer.

5. Use your graph to find the equivalent Fahrenheit temperature for −8°C.

   A 18°F

   B 28°F

   C 42°F

   D 46°F

6. What Celsius temperature is equivalent to −58°F?

   F −50°C   **H** 50°C

   G 14.4°C   **J** −40°C

7. Which is not a solution for the equation in Exercise 1?

   A (100, 212)   **C** (−40, 104)

   B (0, 32)   **D** (60, 140)

Holt Mathematics

Problem Solving

LESSON 12-1 Introduction to Probability

Floods are categorized by their probability of occurrence. For example, a flood categorized as a 20-year flood means it has a 1 in 20 chance of occurring in any given year. Complete the flood probability chart below. Then use it to answer the questions. Write answers in simplest form.

Flood Probabilities of Occurrence

	Category	Probability Fraction	Probability Decimal	Probability Percent
1.	2-year flood			
2.	5-year flood			
3.	10-year flood			
4.	50-year flood			
5.	100-year flood			

6. Which flood category in the table is the most likely to occur in a given year? The least likely?

_____

_____

7. Following the naming system in the table, what category name would you use for a flood that is certain to occur in any given year?

A a 1-week flood

B a 1-month flood

C a 1-year flood

D a 3-year flood

8. The Yukon River in Alaska had a 100-year flood in 1992. Does this mean that another 100-year flood could not occur on the Yukon River until 2092? Explain.

_____

_____

9. The Mississippi River system had a rare 500-year flood in 1993. What is the percent of probability that another 500-year flood will occur on the Mississippi River system next year?

F 2%

G 0.2%

H 0.02%

J 0.002%

Holt Mathematics

LESSON 12-2

Problem Solving

Experimental Probability

Write the correct answer. Write answers in simplest form.

1. Brandy tossed a fair coin several times. She recorded the result of each toss in this table. What is the experimental probability that Brandy's next toss will land heads up?

Heads Up	ЖЖ ЖЖ ЖЖ I
Tails Up	ЖЖ ЖЖ IIII

2. In this table, Charles recorded the gender of each person who shopped at his store this morning. What is the experimental probability that his next customer will be a woman?

Male	ЖЖ ЖЖ ЖЖ ЖЖ II
Female	ЖЖ ЖЖ ЖЖ III

3. Nita packed 4 pairs of shorts for her beach vacation—a blue pair, a white pair, a denim pair, and a black pair. Without looking, she pulls out the blue pair from her suitcase. What is the outcome?

4. Mick rolled two number cubes at the same time. Each cube is numbered 1 through 6. The cubes showed a sum of 7. What is the outcome for this experiment?

Abdul recorded the number of free throws his favorite basketball player made in each of 24 games. He organized his results in this frequency table. Circle the letter of the correct answer.

Free Throws Made	0	1	2	3	4
Frequency	1	4	7	9	3

5. What is the experimental probability that this player will make 1 free throw in the next game?

A $\frac{1}{24}$

B $\frac{1}{6}$

C $\frac{7}{24}$

D $\frac{1}{8}$

6. Based on Abdul's experiment, how many free throws will this player most likely make in any given game?

F 3

G 4

H 0

J 2

Holt Mathematics

Name _____ Date _____ Class _____

Problem Solving

Counting Methods and Sample Spaces

Write the correct answer.

1. Computer spreadsheet programs use letter-number combinations to name cells. How many different cells can a spreadsheet have where its name has 1 English letter followed by 1 digit?

_____

2. An airline has five different flights to San Francisco today. Each flight offers first-class or coach seats. From how many different tickets to San Francisco can you choose today?

_____

3. On Friday, the school cafeteria is serving pizza, hamburgers, chicken, milk, chocolate milk, and juice. From how many different meal-drink combinations can you choose?

_____

4. Tanya packed 4 T-shirts, 6 pairs of shorts, and 2 pairs of shoes for her vacation. How many different short-shirt-shoes outfit combinations can she wear?

_____

Circle the letter of the correct answer.

5. There are 4 people at a meeting. Every person shakes hands with each other person once. How many handshakes are done in all?

   A 16 handshakes

   B 12 handshakes

   C 8 handshakes

   D 6 handshakes

6. There are 3,628,800 different ways to arrange the digits 0 through 9! How many different ways can you arrange the digits 1, 2, and 3?

   F 4 different ways

   G 6 different ways

   H 7 different ways

   J 9 different ways

7. A spinner has 6 equal sections labeled *A, B, C, D, E,* and *F.* A second spinner has 5 equal sections colored red, blue, green, yellow, and black. If you spin both spinners at the same time, how many different possible outcomes are there?

   A 5   C 11

   B 6   D 30

8. How many different ways can you get from point *A* to point *G*?

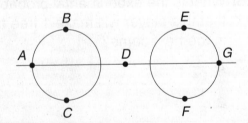

   F 4   H 5

   G 9   J 12

Holt Mathematics

LESSON 12-4 Problem Solving
Theoretical Probability

Each time a letter is drawn, it is returned to the bag. Write the correct answer. Write answers in simplest form.

1. At the beginning of a game, each player picks letter tiles from a bag without looking. What is the probability that a player will pick a blank tile?

   _____

2. Which letter are you most likely to pick from the bag? Write this probability as a fraction, decimal, and percent.

   _____

3. Which letters are you least likely to pick from the bag? What is the probability that you will pick any one of those letters? Write this probability as a fraction, decimal, and percent.

   _____

Numbers of Tiles for Each Letter

Letter	Tiles	Letter	Tiles
A	9	O	8
B	2	P	2
C	2	Q	1
D	4	R	6
E	12	S	4
F	2	T	6
G	3	U	4
H	2	V	2
I	9	W	2
J	1	X	1
K	1	Y	2
L	4	Z	1
M	2	BLANK	2
N	6		

Circle the letter of the correct answer.

4. The probability of randomly picking a letter is $\frac{3}{50}$. What could that letter possibly be?

   A E
   B G
   C N, R, or T
   D V, W, or Y

5. The probability of randomly picking a letter is $\frac{1}{25}$. What could that letter possibly be?

   F A
   G B
   H C, F, H, or M
   J D, L, S, or U

6. What is the probability that you will select a vowel tile (including Y) from the bag?

   A $\frac{9}{100}$ C $\frac{11}{25}$
   B $\frac{26}{49}$ D $\frac{21}{50}$

7. Most words with a Q must also have a U. What is the probability that you will select a U?

   F $\frac{1}{100}$ H $\frac{1}{20}$
   G $\frac{1}{25}$ D $\frac{1}{300}$

Holt Mathematics

Problem Solving

LESSON 12-5 *Compound Events*

You have two decks of playing cards. You draw one card from each deck at the same time. Write the correct answer.

1. What is the probability that you will draw a black card from Deck 1 and a red card from Deck 2?

   _____

Standard Deck of Playing Cards

Suit	Color	Number
Spades	Black	13
Hearts	Red	13
Clubs	Black	13
Diamonds	Red	13

2. What is the probability that you will draw a club card from both decks?

   _____

3. What is the probability that you will draw a heart from Deck 1 and a black card from Deck 2?

   _____

You roll two standard number cubes at the same time. Circle the letter of the correct answer.

4. What is the probability that you roll doubles, or the same two numbers?

   A $\frac{1}{2}$

   B $\frac{1}{3}$

   C $\frac{1}{6}$

   D $\frac{1}{12}$

5. What is the probability of rolling a sum less than 6?

   F $\frac{5}{18}$

   G $\frac{1}{6}$

   H $\frac{1}{9}$

   J $\frac{1}{18}$

6. Which sums are you least likely to get? What is the probability of rolling either of those sums?

   A 2 or 3; $\frac{1}{12}$

   B 2 or 4; $\frac{1}{9}$

   B 2 or 6; $\frac{1}{6}$

   D 2 or 12; $\frac{1}{18}$

7. Which sum are you most likely to get? What is the probability of rolling that sum?

   F 7; $\frac{1}{6}$

   G 8; $\frac{1}{9}$

   H 9; $\frac{1}{9}$

   J 10; $\frac{1}{12}$

Holt Mathematics

Name _____ Date _____ Class _____

Problem Solving
Making Predictions

Write the correct answer.

U.S. Public High School Graduation Rates, Top 5 States

State	Number of Students	Percent that Graduate
Iowa	497,301	83.2%
Minnesota	854,034	84.7%
Nebraska	288,261	87.9%
North Dakota	112,751	84.5%
Utah	480,255	83.7%

1. In which state are students most likely to graduate from public high school? About how many of the students who are enrolled in that state now do you predict will graduate?

2. About how many students enrolled in North Dakota public high schools now do you predict will graduate?

Circle the letter of the correct answer.

3. About how many students enrolled in Minnesota public high schools now do you predict will graduate?

   A about 717,389 students

   B about 723,367 students

   C about 743,010 students

   D about 7,233,667 students

4. About how many more students in public high schools do you predict will graduate in Iowa than in Utah?

   F about 413,754 more students

   G about 401,973 more students

   H about 11,781 more students

   J about 1,781 more students

5. The total U.S. high school graduation rate is 68.1%. There are 48,857,321 students enrolled in public schools. About how many of those students do you predict will graduate?

   A about 332 million students

   B about 20 million students

   C about 33 million students

   D about 16 million students

6. About 11% of all students in the U.S. are enrolled in private schools. There are more than 48 million students in the U.S. About how many do you predict will go to private schools?

   F about 5,280,000 students

   G about 6 million students

   H about 52,800 students

   J about 528,000 students

Holt Mathematics